THE GOATKEEPER'S VETERINARY BOOK

To
Louise

THE GOATKEEPER'S VETERINARY BOOK

by

PETER DUNN, B.V.Sc., M.R.C.V.S.

Illustrations by Louise Dunn

FARMING PRESS LIMITED

WHARFEDALE ROAD, IPSWICH, SUFFOLK

First published 1982

Second impression 1983

ISBN 0 85236 121 1

Printed in Great Britain by John G Eccles Printers Ltd, Inverness.

CONTENTS

ILLUSTRATIONS

FIGURES

PREFACE

The world population of goats is 450 million. Their products feed and clothe a substantial portion of the human population. Most goats are owned by the poorer members of society and there has been tremendous prejudice against the goat by most organisations concerned with agricultural development. Poor people, often illiterate, have made no records and agricultural ministries tend to omit the goat from their surveys. As a consequence, knowledge of the diseases of goats is limited. Fortunately the situation is now changing and organisations are beginning to promote goat projects.

The treatment of goats in the United Kingdom has been rather similar. The Ministry of Agriculture, Fisheries and Food, for example, has almost completely ignored the animal, although more recently there have been welcome signs of interest developing. In the past many veterinary surgeons have had little experience with goats, and as a consequence goatkeepers have tended to use them as a last resort. Unfortunately, this has also meant that the veterinary surgeon sees very few goats and therefore never develops any expertise with them. Over the past few years the veterinary profession has taken an active interest in the goat and now a Goat Veterinary Society has been formed. Hopefully, when goatkeepers realise that more can be done for their charges they will use their vet more wisely.

It is against this background that this book is written. I have attempted to draw together all the relevant information which is widely scattered in books and journals and to present it in a readily usable form for the goatkeeper. Drug details are given for the benefit of veterinary surgeons seeking information about drugs known to have been used on goats. Dosage rates have been included because these are frequently not given by the manufacturers. Throughout this book the reader will notice that I have reduced the emphasis on the role of the micro-organisms (bacteria and viruses, for example) that was so characteristic of older works. Instead I have tried to place more emphasis upon the factors contributing to disease. The reason for this should become evident. Most of these micro-organisms are present all the time; it is changes in the other factors, such as diet and housing, which precipitate the disease. These are the factors that you the goatkeeper can attempt to modify in favour of the goat in order to avoid disease.

Shinfield, Reading,
March 1982   PETER DUNN

ACKNOWLEDGEMENTS

I would like to thank the following people for their help in the preparation of the book: Heidi, Jane, Linda and Mary for help with the typing. For their constructive criticism of the text Margaret Jones M.R.C.V.S., and Alan Slater M.R.C.V.S. deserve a special thankyou, as does my sister Hilary Kewin for taking the trouble to read through the manuscript. Sue Clarke deserves a mention for her valuable advice with artwork. I am particularly grateful to Alan Mowlem of N.I.R.D. for his help with photographs. I am also grateful to the following for their help with various aspects of photography:

The Photographic Department of Reading University
Dr Alex Donaldson, M.V.B., M.A. and
Jennifer Ryder I.V.R.I., Pirbright
Dr A. H. Andrews R.V.C., London
Peter Jackson F.R.C.V.S., Cambridge Veterinary School
The Photographic Department, N.I.R.D., Shinfield.

An acknowledgement is due to Professor R. V. Short for allowing us to adapt one of his illustrations of the goat's udder, and I am grateful to Phil Joyce for reading the proofs.

Finally a special thankyou to my wife, Louise, for all her help and encouragement throughout the preparation of this book.

GENERAL INFORMATION

PHYSIOLOGICAL DATA

Great care should be taken in interpreting the following 'normal' measurements because they are extremely variable. These measurements should be taken when the goat is rested. It is no use counting the goat's breathing, for example, after chasing it round a field.

Body temperature	39-40.5°C
Pulse	77-89 beats per minute
Respiration	15-25 per minute
Age at puberty	4-5 months (male and female)
Season	September to February (approx.)
Duration of oestrus (heat)	12-48 hours
Length between heats	19-21 days (within the season)
Pregnancy	146-154 days (average 150)

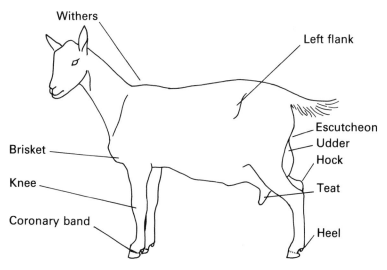

Diagram 1.1 POINTS OF THE GOAT

Chapter 1

PRINCIPLES OF HEALTH AND MANAGEMENT

HEALTH AND DISEASE

The emphasis of this book is on keeping goats healthy and free from disease. The word 'disease' describes precisely what it is: *dis-ease*, the state when an animal is not at ease with its surroundings. Health is more difficult to define but it is the state when an animal is at ease with its surroundings. Common usage has resulted in the word disease being equated with infections, whereas the situation is, in fact, far more complex.

THE WORLD THE GOAT LIVES IN

The factors that decide whether or not an animal succumbs to a disease condition are many and varied. They can be conveniently categorised, however, as:

1. The goat's own peculiarities;
2. The surroundings in which we keep it;
3. The infectious agents in that environment.

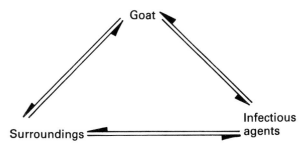

1.2 Factors determining health and disease

All three groups of factors influence each other, as represented in fig. 1.2. One could consider health to be the state of affairs when a balance is struck between these three factors. Disease on the other

15

hand occurs when there is an imbalance of these forces.

THE GOAT'S DEFENCE MECHANISMS AGAINST DISEASE

There are three types of defence mechanism:
1. General defence and the normal functioning of the goat;
2. Cellular defence system;
3. Humoral (antibody) system.

The General Defence System
Examples of the general defence system are the skin, which covers the body and protects it from invasion; the acidity of the true stomach, which destroys some micro-organisms taken in with the food; and also the upper respiratory tract. The latter contains the turbinate bones in the nose which trap particles of dust before they enter the air passages. This function is assisted by the mucus lining the whole of the air-conducting part of the system.

The Cellular Defence System
Special cells (microphages and macrophages) found in the tissues and circulating bloodstream are capable of destroying micro-organisms that have entered the goat's body. They accomplish this by phagocytosis, literally eating up and digesting the invader.

The Humoral (Antibody) Defence System
Specific antibodies capable of inactivating bacteria and viruses are found in abundance in the goat's tissues. Antibody is also produced against worm parasites but it is generally fairly poor and short-lived. The production of antibody always follows exposure to a specific micro-organism. Repeated exposure increases the quantity produced. The process is mimicked in vaccination when minute doses of inactivated micro-organism (or part of the micro-organism) are given to the goat to stimulate its production of antibody. Commonly used vaccines for goats are against tetanus and enterotoxaemia.

SUSCEPTIBILITY AND RESISTANCE

All goats are either susceptible or resistant to diseases with all the grades in between. When considering infectious diseases, the key to avoiding problems is to know which goats are susceptible and which are immune. Kids are susceptible to most diseases because they have not developed antibody against them. If they have received adequate colostrum, however, they are protected against most of the diseases to which the dam is immune. This maternal immunity rapidly declines over the first few weeks of life and the kid must then start to manufacture its own antibodies.

Gradual exposure to most disease agents can be considered to be akin to a natural vaccination process. Goatkeepers must be aware that a kid's resistance may be very different from an adult's, and they must plan accordingly. It would be unwise, for example, to mix a large group of adult goats in the same house as a large group of kids. The kids would be very prone to respiratory disease. Similarly when at pasture the kid is very much more susceptible to worm parasites than is its dam.

STRESS AND DISEASE

The word 'stress' is frequently used in this modern age to refer to the 'stress' of business life or of living in a big city. We all think we know what stress is, but when does it become involved in goat disease? One famous biologist (Seleye) put forward a theory to explain the exact mechanism of stress. His view was that any detrimental influence on an animal, such as a goat, is a stressor. These influences are legion and diverse, but include factors such as bullying by other goats, long journeys and climatic factors such as heat. The crux of the matter is that they are all transmitted by the goat's brain, and have a common effect on the physiology (or internal workings) of the goat. This effect is thought to be the release of excessive quantities of hormone from the adrenal cortex. These hormones reduce the goat's defence mechanisms and therefore render it more susceptible to infectious diseases. (see fig. 1.3.)

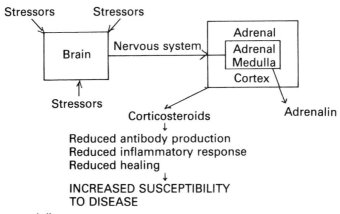

1.3 Stress and disease

Good Stockmanship

Every-day dealings with goats, or other animals for that matter, should be directed to reducing stress, which means simply that common sense in feeding, housing and management are needed. Avoid sudden changes in diet, avoid unsuitable floors that lead to goats slipping and injuring themselves. It is all the little points that distinguish good goatkeepers from bad ones, and this manifests itself in the health of their animals.

The Importance of Good Husbandry

Over the last decade research work on the dairy cow has shown that poor husbandry is the most important factor in numerous common disease conditions such as mastitis, acetonaemia, infertility and lameness. All these points are equally applicable to goat husbandry, and remember that avoiding disease by good husbandry saves money on veterinary fees.

Vaccines and Vaccination

Vaccines are an economical method of protecting goats against some diseases. Used in accordance with the manufacturer's instructions they are normally safe and effective. They are not magic potions and, for example, the goats must be adequately fed if they are to be able to respond to them. Likewise, the ability of a goat to defend itself wanes as old age approaches.

Some vaccines (such as enterotoxaemia vaccine) contain adjuvants. These are compounds such as aluminium hydroxide which enhance the immune response by prolonging the release of vaccine from the injection site. They tend to cause 'lumps' at the site of injection, which may persist for months. Injections are therefore best administered inside the foreleg or at other sites where the lumps are not visible. The following points should be observed when using vaccines.

1. Carefully follow instructions on use such as 'keep cool' (if directed) in a refrigerator at +4°C.
2. Be 'clean' when administering them; swab the site with spirit or alcohol. If multidose containers are used avoid contaminating the vaccine with a dirty needle.
   Always use two needles, one in the container and one for injecting the goat.
3. Check the expiry date in case the vaccine is 'out of date'.
4. Avoid injecting or pricking yourself. This is especially important if using vaccine against orf.

5. Keep a record of dates when vaccines were given.
6. Check the minimum age of receiving the vaccine; for instance, for enterotoxaemia this is ten weeks for kids born to vaccinated does, one week for kids born to unvaccinated does.

SIGNS OF HEALTH AND ILL HEALTH

Recognising that goats are sick is easy when something very obvious such as bloat occurs. Being able to spot the more subtle indications of ill health is rewarding because treatment can be initiated quickly. This may mean fewer deaths or fewer animals becoming ill because action is taken in time. The signs that the goatkeeper should monitor every day are listed below:

- *General attitude*: alert, inquisitive.
- *Appetite*: the goat should be interested in food at almost any time.
- *Cudding*: at certain times of the day your goat should chew the cud.
- *The eyes and nose*: bright eyes, no discharges and a cool, dry nose.
- *The coat*: clean and glossy.
- *The droppings*: firm and pelleted.
- *The urine*: light brown, no blood in it.
- *Breathing*: regular and unlaboured.
- *The gait*: steady, all feet taking weight as the goat walks.
- *Milk yield*: sudden changes should be noted.
- *The milk*: changes such as clots or blood are abnormal.

BE OBSERVANT: In order to notice deviation from normality one must be very familiar with the normal situation. All goatkeepers should be unconsciously noticing little points about their stock as they tend them. Anything abnormal should be carefully noted in case it turns out to be significant.

NURSING SICK GOATS

Goats have a reputation for being poor patients in that they seem to give up the will to live. Therefore, it is all the more important that goatkeepers be diligent in nursing their charges. From observing sick animals on different farms and smallholdings it has become clear to me that some people have a gift for nursing while others do not. The veterinary surgeon can only prescribe and treat sick animals, the owners must carry out the important task of nursing.

Firstly, like humans, animals respond to their surroundings, so put them in a pleasant, airy building, with light and possibly access to the outside. The floor should be well strawed up. Avoid shutting them up in a north-facing 'Black Hole of Calcutta'. Water should always be on offer, but not food. If a goat is not eating offer it a range of tit-bits from time to time, but try to avoid leaving it there for long periods. Offer only small quantities and choose appetising or unusual foodstuffs. For example, a sick goat will often nibble at apple tree prunings or ivy leaves.

The approach one takes obviously depends upon the condition being nursed. Animals with severe pneumonia, for example, would benefit from having their water supply at head height. Never be afraid to let a sick animal outside if it shows any desire to go out. A nibble at greenery may do it the world of good.

Bed Sores
Goats that are down for long periods soon develop sore patches of skin where they are in contact with bedding. Keep the bedding as dry as possible and reposition the goat twice a day. Encourage it to stand occasionally. Goats should never be allowed to lie flat on their sides for long periods but should be encouraged to sit on their brisket (propped if necessary), to avoid risk of bloat.

The Animal's State of Mind
Some goats that have been 'down' for some days become convinced that they cannot get up. Frequently they will get up if they are encouraged and if the goatkeeper assists by supporting their rear quarters.

General Treatments
Specific treatments have been dealt with in other parts of this book but a few general comments are applicable here. Fluid therapy is increasingly used for many conditions with good results, especially when the goat will not eat. This means that fluids and salts are administered either orally or by injection. Many veterinary surgeons find injections such as glucose saline or Duphalyte (Duphar Ltd) useful in these instances.

WHEN TO SEEK HELP FROM YOUR VETERINARY SURGEON

It is extremely difficult to give any general rules as to when a goatkeeper should seek the help of his or her vet because it depends partly on the goatkeeper's experience. By looking up the relevant section in this book you will hopefully receive some guidance, but that often presupposes that you know what you are dealing with. Most vets are prepared to answer simple questions on the phone, but remember that they do not earn any money during these conversations and are not getting their calls done.

PLATE 1.1
Taking the temperature of a kid

SEEK ATTENTION QUICKLY IN THE FOLLOWING CIRCUMSTANCES:

Broken limbs
Severe bleeding
Goats with nervous symptoms
Suspected poisoning
Protracted birth of kids
Prolapse of the womb following kidding.

Delays can be detrimental; give your vet plenty of warning so that in the case of a difficult kidding, for instance, he has a chance to deliver live kids.

Seek help less urgently with goats that refuse food for longer than eight hours.

BUYING STOCK

Your Foundation Goats

A few points should be considered when you are buying your first goats. If your premises has had no goats on it for over two years it should be free from most problems of goats. This is a wonderful opportunity to start off with a clean bill of health. If the ground that you intend to keep your goats on has carried sheep, then you can be less sure of a clean start because goats and sheep have many diseases in common.

Examine your prospective purchase carefully, if possible in daylight. Pay particular attention to:

• Bright eyes.
• Alert expression.
• Absence of discharge from the eyes and the nose.
• A goat that stands well and square; not a goat that keeps wanting to lie down (is it lame?)
• Look for cud chewing, but avoid a goat that 'drops its cud' or drools saliva (adult).
• Avoid goats that breath excessively fast at rest, or goats with diarrhoea.
• Look for a clean healthy-looking coat which is free of parasites.
• The udder attachment should be over a large area, with average-length teats (not too long). Feel the udder which should be soft and free from hard lumps (adults). Check that there are only two teats.
• Check for other obvious problems such as a hernia.
• Examine the teeth in order to assess the goat's age.

General

Enquire from the owner if and when the goat was vaccinated, and obtain the date of the last worm treatment. Ask the vendor if any of their goats have Johne's disease; if the answer is 'yes' then avoid purchasing goats from that source (*See* JOHNE'S DISEASE, page 106). Find out from the owner how the animal has been kept for the previous six months, whether it has been out at pasture, inside all the time, or whatever. This is especially necessary when purchasing young kids, and you should also be sure to obtain a clear picture of the diet of young kids. Look at any other animals on the premises while you are there and note any signs of ill health that might affect your purchase. You may decide to ask your vet to give your prospective goat a health examination, and it could be arranged that you will return the animal if your vet advises you against buying it.

When You Get Them Home
Ensure the goats have clean fresh water offered in a receptacle that *cannot* be fouled by droppings. If possible, house them for the first night, especially if you are going to worm them; this avoids the goats contaminating your new pasture with worm eggs and larvae. Worm the goats if they have been at pasture in the previous six months. Offer some hay to goats of all ages, but avoid over-feeding. If the goat is lactating, feed her exactly what the previous owner fed her, but certainly no more. Non-lactating goats are best fed smaller quantities than normal, especially after a long journey. Young kids below six weeks, fed with milk or milk replacer, should have their ration halved for the first feed. Water can be offered to make up the volume after they have drunk their half-quota. In general you will get few problems if goats are under-fed for the first few days, but many more if they are over-fed. If your goat has not been vaccinated against enterotoxaemia contact your veterinary surgeon to have this done.

GOATKEEPERS AND THE LAW

Notifiable Diseases
Certain diseases affecting goats are listed as being 'notifiable' by the Ministry of Agriculture, Fisheries and Food. They tend to be infectious and contagious diseases that affect all ruminants, and in order to control them it is necessary to have co-operation between the owners and the ministry. Diseases falling into this category include anthrax and foot-and-mouth disease. If goatkeepers suspect these diseases they should contact their veterinary surgeon, the Divisional Veterinary Officer or the police. Should these diseases be confirmed the ministry has wide-ranging powers to deal with the situation.

Movement Records
In order to be able to trace possible contacts in the event of an outbreak of Notifiable Disease, the MAFF stipulates that goatkeepers should keep a record of the movements of their animals. Records should be kept of does being taken to the buck or to shows or any other movements.

PREVENTIVE MEDICINE

Obviously, the aim of all goatkeepers is to promote good health and avoid disease in their animals. Using vaccines, worm preparations and suchlike are some of the ways of achieving this goal. Disease can be prevented, of course, by practising the best kind of husbandry, and advice can be obtained on this. Some goatkeepers with large herds arrange a regular visit by their veterinary surgeon in order to discuss this very point. Such a visit is aimed at going over the routines of the farm and identifying areas where advice is needed. It may be, for example, that a ten-minute discussion on mastitis control can save you milk that would otherwise have been lost through this costly disease.

PLATE 1.2
Given shelter, goats survive the harshest weather

THE GOATKEEPER'S VETERINARY CUPBOARD

1 clinical thermometer
Scissors, round-ended, curved 6 inch
Sharp knife
Hoof knife
1 pair 6-inch forceps
1 lamb reviver (stomach tube)
A torch
Cotton wool
Several bandages 5 cm × 5 cm
Box of adhesive plasters
Soap
A clean towel

Antiseptic/disinfectant such as Dettol
Antiseptic powder
Surgical spirit
Vaseline petroleum jelly
Udder cream
Bicarbonate of soda
500 g Epsom salts
1 tin of treacle
Electrolyte powder
250 g glucose powder
500 ml colostrum in plastic bottle (deep frozen).

Chapter 2

PROBLEMS OF KIDS

Before going into detailed descriptions of specific problems, it is important to familiarise the reader with some aspects of the anatomy and functioning of the unweaned kid.

The main requirements for the newborn kid are shelter, warmth and milk from its dam (colostrum). Kids are frequently born at unfavourable times of the year (springtime) when ambient temperatures are low and the risk of chilling is high, and fig. 2.1 stresses the importance of these factors.

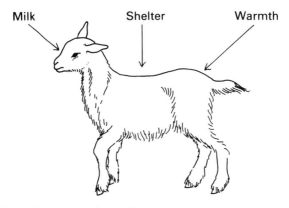

| Milk | Shelter | Warmth |

2.1 Basic requirements of the newborn kid

The provision of these requirements gives the kid a fighting chance to live; failure to provide any one requirement can lead to death of the kid. After birth the kid begins to use up its stores of energy and, unless that store is replenished by sucking milk, the kid becomes chilled. Cold weather accelerates the consumption of the kid's energy store, making it more urgent that the kid takes milk.

DIGESTION IN THE UNWEANED KID

The digestive process of the unweaned kid differs markedly from that of the adult goat (*see* page 63). Whereas the unweaned kid

25

resembles a single-stomached animal such as the dog or cat, adult goats have a complex series of stomachs designed for the digestion of grass and herbage. Although the kid is born with a complete set of four stomachs, a special mechanism operates to by-pass all but the true stomach or abomasum. This mechanism is the oesophageal groove.

The Oesophageal Groove

The oesophageal groove connects the oesophagus to the abomasum as shown in fig. 2.2.

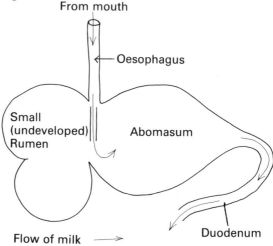

2.2 The position of the oesophageal groove

When the kid sucks a reflex occurs which results in the groove forming a tube (*see* fig. 2.3).

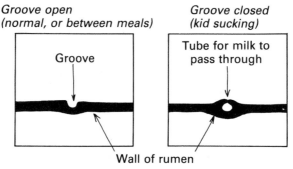

2.3 The oesophageal groove (in section)

Milk can then pass directly down the oesophagus to the abomasum. This mechanism avoids milk stagnating in the undeveloped rumen.

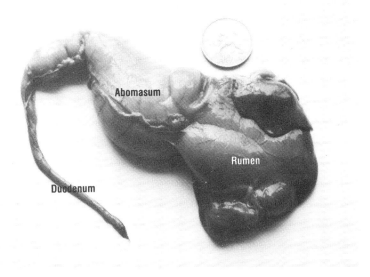

PLATE 2.1
The stomachs and duodenum of a 24-hour-old kid showing the undeveloped rumen

The Oesophageal Groove Closure Reflex

It is important for goatkeepers to understand the stimuli which evoke the groove closure, and therefore avoid digestive upsets in young kids. Whilst it is widely accepted that many substances such as milk or some salts can cause the groove to close, the 'psychological' influence may be little appreciated. Closure of the groove can occur in response to 'conditioning', even in the adult. Thus if the kid is 'conditioned' or accustomed to drinking milk soon after it sees a milk bottle, then very soon just the sight of the milk bottle causes groove closure. Similarly if the kid associates drinking milk with the rattle of certain buckets, this noise alone will stimulate groove closure after a short time. The importance of this in keeping kids healthy is obvious. Always be consistent in feeding routines. This will ensure that groove closure has occurred just before feeding, simply because the signs received by the kid are consistent each feeding time.

THE NEWBORN KID

CHILLING AT BIRTH (HYPOTHERMIA) AND STARVATION

As I have stressed, it is essential that the newborn kid takes in milk as soon after birth as possible. The colder the surrounding temperature, the more urgent this first drink becomes. If milk is not drunk in time, chilling or hypothermia results, because no energy is available for the kid.

Symptoms
Shivering is the first symptom, as the kid attempts to release heat as a by-product of muscular activity. The kid walks around stiffly and aimlessly, until eventually it becomes quiet and comatose. If left it will quickly die, but if treated a large proportion of cases will respond. Never give up with such a kid until you have tried the following course of action.

Treatment

This consists of:

- Drying the coat by rubbing with a warm towel;
- Putting the kid into a warm place, such as by the kitchen stove or a radiator;
- Giving it instant energy (usually a glucose solution).

Giving Glucose
Trickling fluids down the mouth of a comatose kid is dangerous, but if all else fails, give it an egg-cupful of warm water containing a teaspoonful of glucose.

By stomach tube. Many shepherds and vets have a small soft rubber tube with a reservoir on the end, which is used to dose glucose solution directly into the stomach (*see* plate 2.2). These are obtainable from many farm supply dealers. The tube is introduced into the mouth over the tongue and gently pushed down the throat. It is easily done in a comatose kid but not so easy with a fully conscious one.

By injection. In some cases your vet may use an injection in order to revive comatose kids, but many can be saved by your own efforts. Once the kid becomes conscious, it should be encouraged to take in colostrum either directly from the doe or from a bottle with a teat on it.

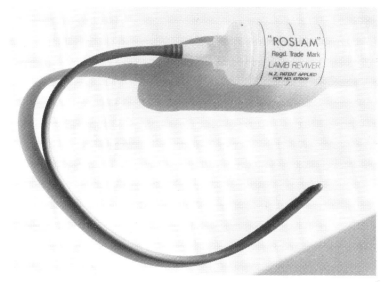

PLATE 2.2
Stomach tube (lamb reviver)

PLATE 2.3
Administration of glucose by stomach tube to a comatose kid

CLEFT PALATE

During the development of kids failure of the formation of the roof of the mouth may occur resulting in a cleft palate (*see* plate 2.4). Problems arise when the kids attempt to suck. Because of the difficulty of forming a vacuum in the mouth, little milk is obtained during suckling. Some will be seen refluxing down the nose and the kid may cough and sneeze in between sucking milk.

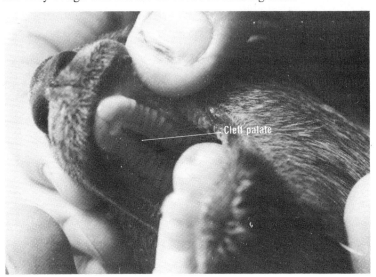

PLATE 2.4
Cleft palate in a newborn kid

Treatment
There is no treatment, the kids are best put down soon after birth.

Prevention
The cause is unknown but vitamin A deficiency in the dam during pregnancy may well be involved.

YOUNG KIDS

THE MAJOR CONSIDERATIONS

Checklist
1. Vaccination of the dam
2. Colostrum intake

3. Hygiene
4. Environment
5. Space allowance
6. Number of kids in the group
7. Source of the kids.

Vaccination of the Dam and Colostrum Intake
Measures taken to avoid problems with kids start way back in pregnancy with vaccination of the doe. Protection for the kid is passed in the mother's first milk (colostrum), when the kid sucks its first meal. Colostrum contains a 'package deal' of antibodies precisely formulated for the environment of the goat. Thus, if possible, goats should be kept in the place or area where they are due to kid for at least fourteen days prior to kidding. This enables the doe to manufacture and deliver to the kid the correct range of antibodies. The next essential step is to ensure that this colostrum is consumed as soon as possible because the kid's intestine can only absorb it for a time limited to about twelve hours.

Hygiene
At birth the kid passes from the sterile womb to the contaminated environment of the outside world. As long as the young kid is only gradually exposed to various disease agents few problems arise. Big problems can occur, however, if hygiene is poor and colonisation of the kid's intestine by bacteria is too rapid. Thus, all utensils and fixtures with which the kid comes into contact should be as clean as possible. In general, hygiene becomes less important as the kid matures, but one must start off carefully.

Environment
Disease is rarely the simple invasion of an animal by microbes; other factors such as the environment play an important part in determining whether or not an infection causes disease. The better the conditions in which we keep our kids, the greater the microbial challenge that they can resist. The requirements for kids are a warm dry bed and good ventilation which is free from draughts. It is important to remember that kids are unable to produce as much body warmth as their parents because the kids' rumens are not functional.

Space Allowance and the Number of Kids in a Group
The space allowance of the kids is important because too little means that they are more heavily challenged by microbial agents.

When large groups of kids are kept together further complications may arise, despite the fact that this space allowance is apparently adequate. Bedding must be replenished frequently, otherwise coccidiosis can occur. Many people are concerned about whether kids should be reared singly or in groups of say four or five. I feel that the benefits arising from kids of a similar age group being together, keeping each other warm and content, outweigh the disadvantages. This assumes of course that the checklist of considerations (see page 30), has been taken into account.

Source of Kids
If kids from different sources are purchased for rearing, problems may arise. As stated earlier, the colostral antibody is for a specific environment. It does not necessarily give the correct protection for the new environment if the kid is moved. Also each kid has its own range of micro-organisms and these may be very different from those of its new pen-mates.

Diarrhoea

The most common disease problems of kids are those where the main symptom is that of diarrhoea (scour). Most of the checklist given on page 30 will have a bearing upon the occurrence of diarrhoea. The types of disease agents involved are many and varied, including *E. coli* bacteria and Rotavirus. Knowledge of the actual agent involved is, in most cases, of less importance than avoiding the conditions which allow them to exert their harmful effects. The reason for stating this is that the agents are normally present around the goat and her kids. They only cause trouble if and when they are allowed to do so.

Some Agents Associated with Diarrhoea in Kids

> *Escherichia coli*
> *Salmonella* spp
> *Clostridium welchii*
> *Eimeria* spp
> *Rotavirus*
> *Herpes virus*

Dietary Scour

Diarrhoea frequently develops simply because changes in feeding have been made or too much food has been offered to the kid. The change in food has the effect of modifying the conditions within the intestines perhaps, for example, by altering the acidity or alkalinity. Micro-organisms such as bacteria grow at different rates in different situations and the changes will favour one type, allowing them to flourish at the expense of others. This results in large amounts of toxin being produced which irritates the gut wall causing it to lose water and salts. This is seen as diarrhoea. Many of these bacteria such as *E. coli* are present in healthy kids and they only get out of hand when conditions change. Preventing this type of scour depends upon avoiding drastic feed changes.

The danger of these upsets is that they predispose to entero-toxaemia (*see* page 35).

(*See also* DIETARY SCOUR IN ADULT GOATS, page 72.)

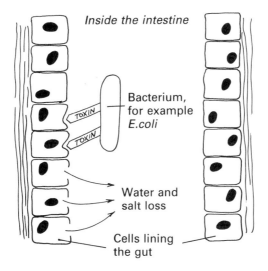

Inside the intestine

Bacterium, for example *E.coli*

TOXIN

TOXIN

Water and salt loss

Cells lining the gut

2.4 Bacterial toxins acting on the intestine to cause loss of fluids

SALMONELLOSIS

Infection by many of the 1600 serotypes or strains of *Salmonella* organisms can occur in goats. Many cases have been reported overseas, but it is seen only rarely in the United Kingdom. Serotypes have included *S. typhimurium*, one which can cause food poisoning in human beings. Infection with *Salmonella* organisms does not necessarily result in disease, there are many healthy goats shedding this germ in their droppings. Trigger factors such as stress can precipitate the disease of Salmonellosis.

PLATE 2.5
Kid with dietary scour

Symptoms
Kids are generally the most severely affected, with symptoms including diarrhoea, high temperature, blood poisoning and, often, death. Because of the acute nature of the disease the diarrhoea is often blood-tinged or even black.

Prevention
Avoiding all the situations which impose stress on the kids is the surest way to stay clear of the disease; among these situations are

irregular feeding, draughts and long journeys. The problem cannot be controlled by vaccination because there are so many different strains.

In adult goats salmonellosis is rare but may occur when goats are exposed to the stressors mentioned above.

Treatment
The severe nature of this disease in kids will dictate that the goatkeeper seeks the assistance of his vet. Treatment must be started early and your vet will probably use antibiotics and fluid replacement therapy. The proportion of deaths among kids affected by this condition can be very high.

Coccidiosis
Diarrhoea is one of the main symptoms of coccidiosis. However, as the problem is most intimately connected with housing conditions I have included it in Chapter 3 (*see* COCCIDIOSIS, page 57).

ENTEROTOXAEMIA

Vaccination of the doe is especially important in avoiding the problem of enterotoxaemia in kids. Enterotoxaemia is a killer disease in herds not using routine clostridial vaccination. The clostridia are a family of bacteria, the most well known member of which causes tetanus (lockjaw) in humans and animals.

Enterotoxaemia can affect goats of all ages, but tends to be more lethal in young kids. The bacteria are commonly found in the soil and most animals will have them in their intestines. It is, therefore, unrealistic to think of keeping goats free from them.

Prevention
It is possible to increase the immunity of the goat to the clostridia. This is successfully achieved by vaccinating the doe during pregnancy and letting her pass the protection on to the kid by way of the colostrum. Avoiding the conditions which allow the organism to proliferate in the intestines and release their toxins is equally important. Sudden changes in the type of food or quantity fed will aid this proliferation. Changes in the kids' feed must therefore be gradual, and up to a week should be taken to change from one type to another. It is also important to avoid engorgement by kids, for example after they have become excessively hungry. All goats should be vaccinated against this disease at least once a year, and preferably twice. The dose should be twice the recommended sheep

dose. When vaccinating, great care should be taken to avoid injecting into the muscle. The correct site is under the skin.

Symptoms
The symptoms of this disease are sudden in their onset and include depression and a drunken appearance. As the disease progresses the animal becomes unable to stand and lies on its side making paddling movements. Very watery diarrhoea may be seen, depending upon the severity of the condition.

Treatment
Treatment of this condition is rarely satisfactory, but your veterinary surgeon may try giving the sick kid specific antisera. It is often more valuable to use the antisera on other members of the group for the purpose of prevention. Some vets report successful treatment using sulphonamides by mouth.

TREATMENT OF DIARRHOEA

The production of loose faeces is generally an indication that the intestinal system has been damaged and, like most damaged organs, requires resting until repair has been carried out. The function of repair is carried out by the body's own system, but ensuring that the intestine is rested requires the goatkeeper to restrict the kids' intake of food.

Generally, restricting the food for one day will alleviate many scour problems. Only small quantities of food must be offered for the subsequent three days, otherwise the good work will be wasted. For unweaned kids, restriction to only half the normal milk intake is recommended, following twenty-four hours of no milk. It is important to offer water, either in a bottle or in a bucket, when milk is restricted. Water is taken more readily by a scouring kid if it is offered warm, with an ounce of glucose per litre and fed through a teat. The kid must be returned to its normal diet gradually, when the diarrhoea has ceased. Kaolin suspension can safely be given to all scouring kids at a dose rate of 1 ml per kg of body weight daily. (An average kid weighs 4 kg at seven days.) Electrolytes (salts), which are extremely useful for restoring kids to health, can be obtained from your veterinary surgeon and are fed by mouth.

Severe cases of diarrhoea, when the faeces resemble dirty water or when blood is observed in the droppings, probably require help from your vet. It is also advisable to contact your veterinary surgeon if the kids are off their food and are looking dull and dejected. In these instances your vet will probably prescribe electrolytes by injection. Some cases of scour in young kids require antibiotic treatment if septicaemia (blood poisoning) is suspected.

SEPTICAEMIA

Septicaemia or blood poisoning can follow a simple case of diarrhoea. If the antibodies that protect the intestine from invasion are overcome, bacteria can cross through into the bloodstream. Kids that have had colostrum would normally be able to defend themselves against such an invasion because in these cases the antibody from the colostrum is present in the bloodstream and acts by 'mopping-up' the invading bacteria. Should the kid have missed its colostrum the blood poisoning rapidly takes a hold and the kid becomes very ill within twelve hours. The symptoms are those of a sick kid with a high temperature and severe diarrhoea. Sometimes the septicaemia develops so rapidly that diarrhoea is not noticed.

Treatment and Prevention
Antibiotics and fluid therapy can save many of the kids but prevention rests upon the simple rule of ensuring that all kids take in colostrum.

COLIC

Colic can affect young kids especially when dietary changes are made. Introducing milk replacer or mixing it up at the wrong concentration can precipitate colic. Spasm and excess gas production results in the bowel giving rise to pain.

Symptoms
The kid is restless, cries out and tends to stand either with its back arched or with its hind feet placed well back. Its nose and mouth may feel cold.

Treatment
In mild cases the pain quickly passes and the kid returns to normal within hours. Severe cases may require drugs to relieve the pain. Isaverin (Merck) and Buscopan (Boehringer) are such relaxants, given by injection.

NAVEL ILL AND JOINT ILL

These diseases are generally caused by dirty environments. Throughout pregnancy the umbilical cord connects the kids to the placenta, enabling the kid to receive nutrient from the doe. Once separated at birth, the cord rapidly dries out and shrivels, leaving a scar of attachment, the navel. Immediately after birth, the fleshy navel is open to infection and it contains blood vessels which, if infected, can result in an infected liver or possibly blood poisoning. Infection may also spread to the joints via the blood, causing 'joint ill'.

Symptoms

The young kid with this condition has a swollen, painful navel which may look red and 'angry'. The kid is commonly off its food and may or may not have swollen joints, if the condition has progressed to 'joint ill'. More usually the joint involvement becomes evident after some weeks.

Treatment

Navel ill is generally treatable, and your vet will probably prescribe antibiotic injections. Joint ill tends to leave badly damaged joints and it may well be better not to embark on a course of treatment; euthanasia may be preferable.

Prevention

The goatkeeper should aim to prevent this problem by providing hygienic conditions at kidding. The kidding box should be well strawed up and the navel could be dressed soon after birth. For dressing navels (i.e. putting antiseptic substances on to the raw cord), sprays, iodine solutions or powders can be used. Your veterinary surgeon will supply you with something suitable.

TETANUS

Infection of open wounds by the bacterium *Clostridium tetani* results in tetanus (lockjaw). It is not common in goats, but it may affect young kids. As with enterotoxaemia, the organisms are present in the soil in a special form known as a 'spore' In this form the bacterium can survive for many years, always being present, ready to infect deep wounds. The germs multiply in the wound releasing toxins which are harmful to the nerves. The toxins travel up the nerve trunk to the spinal cord and brain, where they cause constant excitability of the nerves. This results in the muscles contracting all the time, and therefore the goat walks stiffly.

Symptoms

A general increase in muscle stiffness is seen causing an unsteady gait. The third eyelid begins to extend over the eye and the animal looks 'anxious'. The symptoms get progressively worse and convulsions may occur. The goat dies because it is unable to breathe.

PLATE 2.6
Tetanus in a kid

Treatment

Goats can be treated with penicillin and antisera, but response is poor.

Prevention

Avoid dirty conditions when wounds such as castration cuts are present and likely to become infected.

Vaccination of all goats is cheap and effective, the vaccine is normally included in the multiple vaccines for enterotoxaemia. Young kids receive protection from their mother's colostrum, providing that she is vaccinated each year during pregnancy.

ENZOOTIC ATAXIA (SWAYBACK)

This condition, causing lack of hind limb co-ordination, has been only rarely recorded in kids; it is similar to 'swayback' in lambs.

Symptoms
The symptoms are of weak kids, unable to rise or 'swaying' at the back end. They are associated with poor development of the nerve fibres. This nerve defect results from a deficiency of copper in the diet of the dam during pregnancy.

Rare Occurrence
Because the individual attention given to goats during pregnancy is generally superior to that given to sheep, this condition is less likely to occur in kids. In the unlikely event of owners suspecting swayback, they should contact their veterinary surgeon to confirm their suspicions. Reports suggest that the condition is incurable. Advice about future prevention should be sought.

Diphtheria (Ulcerative Stomatitis)

This problem occurs very rarely in young goats. It is essentially a disease of dirty conditions and is generally seen in housed kids.

The organism most commonly isolated from infected kids is *Fusiformis necrophorus*. This germ is universally present around animals and infection enters through cuts. It is thought that rough food and teething can be responsible for cuts in the mouth, and in one outbreak all the affected goats had diphtheria in either the mouth, tongue or throat.

Symptoms
These will depend upon the area involved but salivation, noisy breathing and smelly breath are common observations.

Treatment
Good response to treatment can be expected from antibacterial drugs such as sulphonamides. Your veterinary surgeon will prescribe something suitable.

Prevention
Hygiene in the rearing of kids is the key to preventing the disease. Cleanliness of feed and water containers is especially important.

ROUTINE PROCEDURES

The Disbudding of Kids

Horned goats are a danger and a nuisance. Most owners of horned kids have the horns removed when they are about one week old. The technique is termed disbudding and it is a fairly simple job when the goat is young. Conversely, removal of the horns of adult goats is a major operation and should only be carried out in the winter when flies are not active. Horn buds can be removed either by using caustic chemicals or by burning. Burning is by far the most satisfactory method.

Procedure

A general anaesthetic is preferable to local anaesthesia. The hair surrounding the horn bud is clipped away, and some suggest the use of cosmetic hair removers to accomplish this step. Using a very hot iron, your veterinary surgeon will burn away the horn bud taking care to burn all the skin around the base.

PLATE 2.7
Clipping hair prior to disbudding

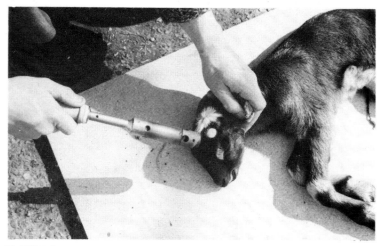

PLATE 2.8
Disbudding the anaesthetised kid

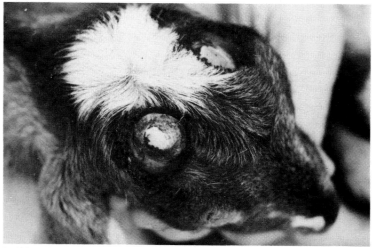

PLATE 2.9
Disbudding completed

CASTRATION OF MALE KIDS

Male kids are neutered for several reasons. Sometimes people wish to rear an animal for pulling a goat-cart or even just to make a non-smelling pet. When rearing male kids for slaughter it is probably best not to castrate them because a better carcase results from an entire male. This may not be possible if the males have to be reared with females because they become sexually active from four months of age. It is also advisable to have them castrated if the rearing period extends through to September — the breeding season. If this happens, the entire males go off their food and do not put on weight, delaying the time for slaughter.

Methods

The testicles must be 'inactivated', either by removing them or by destroying them where they lie. Methods 1 and 2 below are normally carried out under general or local anaesthesia.

1. Removal Using a Knife

The scrotum is incised, the testicles exposed and pulled out. The spermatic cord and the wound are dressed with antibiotic powder or cream.

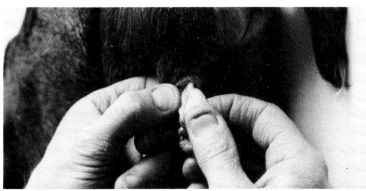

PLATE 2.10
Surgical castration

2. The Burdizzo

Using a special instrument (*see* plate 2.11) the spermatic cord and blood vessels are squeezed between two metal jaws. The effect is to prevent blood reaching the testicles so that they gradually wither away and die. After several weeks the testicles should be re-examined in order to determine whether or not the operation was successful. If the method worked the testicles should be small and hard.

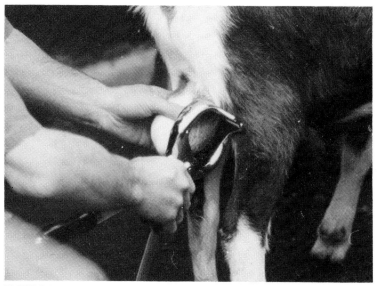

PLATE 2.11
Castration using the Burdizzo method

3. The Rubber Ring Method

A similar technique to (2) is to place a special elastic band around the neck of the scrotum, thus restricting the blood supply to both the testicles and the scrotum. The withered scrotum and testicles eventually drop off about two weeks later. Though effective, this method gives rise to considerable discomfort to the kid and, in the view of many veterinary surgeons is inhumane.

EUTHANASIA OF KIDS

Surplus male kids can be a problem to goatkeepers with limited accommodation. Fortunately, there is an increasing demand for kid meat and there are now outlets for people wishing to dispose of these young kids. Alternatively, your veterinary surgeon will 'put them to sleep' for you, using injections.

Chapter 3

DISEASES OF HOUSED GOATS
RESPIRATORY DISEASES

WHAT IS A RESPIRATORY DISEASE?

A respiratory disease is any condition affecting the breathing apparatus of the goat. This includes the nose, trachea (windpipe), bronchi, and lungs. It may also involve the pleura or membranes which surround the lungs. The respiratory system of the goat is illustrated in fig. 3.1.

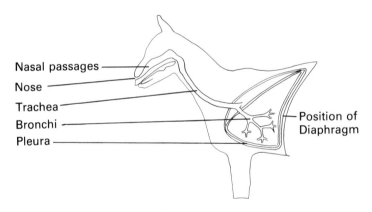

3.1 Breathing apparatus of the goat

BREATHING IN THE HEALTHY GOAT

There is one substance that is present in abundance on this planet, and that is fresh air. It is because of this abundant supply of air that the majority of us keep healthy for most of the time. Animals breathe in order to take oxygen into their lungs and to allow waste gas (carbon dioxide) out. The oxygen is vital for all the tissues of the body. The lungs are the organs that bring about the exchange of oxygen from the air into our blood, and the exhalation of carbon dioxide.

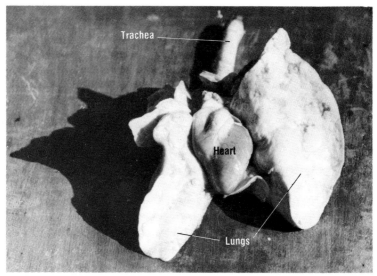

PLATE 3.1
Lungs and heart of a goat

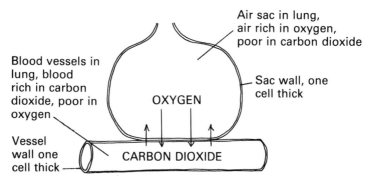

3.2 Gaseous exchange in the lung, showing the movement of gases

Fig. 3.2 illustrates the basic function of the lungs, gaseous exchange. When the goat inhales and exhales, water droplets and micro-organisms are also passing into and out of the goat's respiratory system. The water arises from the moist lung surface and the micro-organisms from all parts of the breathing apparatus such as the trachea, bronchioles and larynx. The breathing apparatus of a healthy goat is colonised by many different types of bacteria, viruses and other micro-organisms. This is normal and natural. In most circumstances the goat remains healthy because it produces anti-body, and because it possesses a complex system which keeps the bacteria and viruses under control.

THE FACTORS PREDISPOSING TO RESPIRATORY DISEASE

These factors are:
1. A shared air space with other goats.
2. Poor ventilation.
3. Age of the housed goats.

A Shared Air Space

When a goat in a field breathes out, the bacteria and viruses which are exhaled in that breath are soon dispersed. The situation inside a small house, however, is very different. The micro-organisms and water vapour are not easily diluted, so they have a profound effect on the air inside, making it humid and laden with many micro-organisms. These micro-organisms are then inhaled by other goats. It must also be remembered that the water vapour will have altered the relative humidity of the air. This is thought to adversely affect the tissues of the respiratory tract, reducing their defensive capabili-ties. The combination of these changes — heavy doses of micro-organisms, and the effect on the lungs and trachea — may precipi-tate respiratory disease.

Ventilation

The principle of ventilation is to ensure that the air within the goat house is changed regularly. This provides fresh air for the goats to breathe and removes the foul air containing the micro-organisms. It is always better to provide more rather than less ventilation. Owners' fears about chilling their goats are unfounded, providing that direct draughts are excluded. Air should circulate freely in the goat house, but you should ensure 'cosy' areas where the goats can be protected from draughts by solid partitions. Remember that the rumen is like a central heating system, but kids do not possess this heat source until they are of weaning age.

Age of the Goats

Another important consideration is the age of the animal in the goat house. Older animals tend to have experienced many bacteria and viruses in the normal course of their lives. Young kids and goatlings are less fortunate. Often the first time they experience a new type of virus is in a goat house where the dose can be high, and they succumb to infection. Thus one must be particularly careful about providing the correct environmental conditions for young goats if respiratory disease is to be avoided.

SIGNS AND SYMPTOMS OF RESPIRATORY DISEASE

The most common type of respiratory disease seen in housed goats is an infection of some part or even the whole of the system. Other less common diseases include allergies, deficiences, cancers and so on. The reasons for the high frequency of infections are the predisposing conditions already described. Of the many types of agents that can be involved, most are viruses and bacteria. For example, one study revealed fifteen different types of bacteria in goat pneumonias. Viruses have the property of being able to invade healthy cells and they tend to commence the disease processes. They damage the tissues, leaving them vulnerable to attack by bacteria and mycoplasma.

The symptoms we see are simply a reflection of the area damaged. For example, an infection of the nose results in a discharge being produced from the nostrils, and similarly an infection of the windpipe tends to stimulate coughing. This cough reflex is unfortunately the best method by which this type of disease passes from one goat to the next. A coughing or sneezing goat produces tiny water droplets containing micro-organisms. These airborne particles are then just the right size to enter the trachea and lungs of any other goats kept in the same building. Once coughing starts, it usually spreads quickly to affect other animals.

MILD UPPER RESPIRATORY TRACT INFECTIONS

These are commonly seen when large groups of young goats are kept together in badly ventilated buildings. The symptoms include coughing, nasal discharge, and raised body temperatures.

Action to be taken
Provide plenty of fresh air and watch carefully for goats going off their food. Any goat that refuses food requires veterinary attention.

It may be necessary to provide areas free from draughts, by using straw bales for goats to shelter behind. These can be renewed frequently. Never be concerned about letting in too much air. Other measures which may be helpful are avoiding dry, dusty food, and reducing the dust from excessively dry bedding materials.

Prevention
There are, unfortunately, no vaccines to assist in the control of these

problems. Methods of prevention are directed at providing the ideal husbandry conditions.

1. Always ensure adequate ventilation. As a rule of thumb, goat houses should be dry and not dripping with condensation. Goats can withstand cold environments, so long as they are not draughty.
2. If possible, avoid mixing young goats (two to eight months) which come from different rearing places.
3. Ensure that all kids receive adequate colostrum immediately after birth, so that they are provided with protective antibodies.
4. Try to keep the bedding dry so that it does not increase the relative humidity of the house.

PNEUMONIA

A better name to describe an infection of the lung would be pneumonitis, but it is traditionally referred to as pneumonia. This can result from a serious upper respiratory tract infection (see above) or possibly from drenching a goat incorrectly. In any event this condition is serious and should receive skilled attention quickly. The symptoms are those of a sick goat which refuses food, probably has an elevated temperature, and coughs and breathes rapidly. Your veterinary surgeon will probably prescribe antibacterial drugs and other aids to remove the fluids from the lungs. Simple first-aid treatment includes the recommendations given under 'Mild respiratory infections'. Water and food should be offered at a suitable height, not on the floor where the animal has to bend down to eat or drink. Both *Pasteurella haemolytica* and *P. multocida* have been isolated from such cases.

Drenching Pneumonia

This follows the accidental pouring of fluids into the windpipe and lungs of a goat undergoing medication. It may also result from force-feeding a kid from a bucket. Symptoms usually develop a day or so following the drenching and the course of the disease can be fatal. Your veterinary surgeon will advise you on the best course of action to take.

OTHER CONDITIONS

Many other conditions manifest themselves by symptoms such as coughing or rapid breathing; an example would be an allergy. The frequency of them occurring, however, is so low that it would be pointless to detail them in a book such as this. Your veterinary surgeon will help you in this situation.

Lungworm

It may happen that housed goats develop lungworm, resulting from an infestation gained while at pasture. *see* LUNGWORM, page 92.

THE FLOOR AND FOOT CONDITIONS

LAMENESS AND HOOF CARE

An important problem of goats kept in houses is that of lameness. Frequently the cause of the lameness is associated with the nature of the floor or bed. When given a choice, goats tend to prefer dry surfaces to walk upon; indeed the habitat of the mountain goat is firm craggy rocks. Dry abrasive surfaces are ideal for the goat's hoof, which continuously produces horn in the anticipation that it will be worn away.

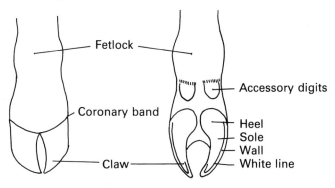

3.3 Anatomy of a goat's foot

Normal Horn Growth

Wall horn is produced from the skin/horn junction of the coronary band, and the horn works its way down until it comes into wear with the ground. Horn is also produced to cover the sole, but this should not be in contact with the ground. In short the weight of the goat should be borne on the edge of the wall horn which surrounds the hoof.

Principles of Hoof Care

Because most goats do not spend enough time walking on dry, abrasive surfaces the owner must pare away the excess horn. If this is not done, incorrect stress is placed upon the hoof, causing cracking of the horn. Stones may also become embedded in the fold of horn. Plates 3.3 and 3.4 show the hoof before and after paring.

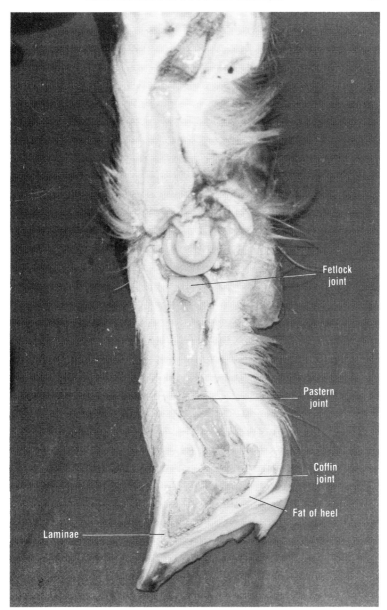

PLATE 3.2
Section through one claw

3.4 The weight-bearing area of the goat's hoof

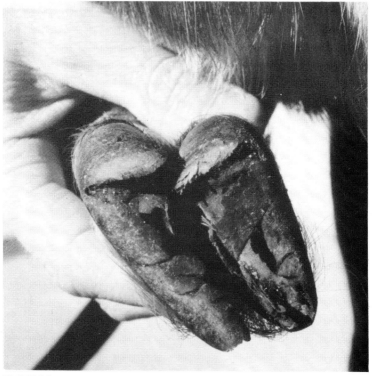

PLATE 3.3
Hoof before paring

The aim of paring is to remove the surplus wall horn that tends to bend under the sole and impede normal action. Paring can be carried out using hoof shears or a hoof knife. A file or a surform plane may also be useful for some cases.

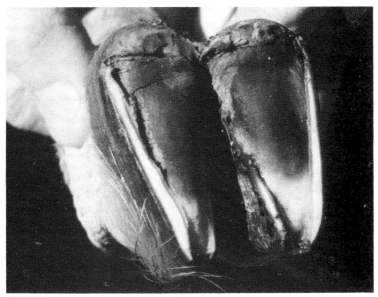

PLATE 3.4
Hoof after paring

Hoof shears

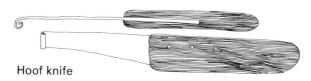

Hoof knife

3.5 Hoof-care equipment

Goats Kept on Deep Bedding

Work with cattle has revealed a great deal of information about the causes of lameness in hoofed animals, which is directly applicable to goats. One important aspect is the dryness or wetness of the floor.

Ideally, as I have suggested, housed goats should be kept on a hard surface. Many systems, however, do not allow this and goats can be successfuly raised on deep straw bedding. In this situation there is no wear on the hoof horn and attention must be paid to paring as described in the preceding section.

Another important aspect is the effect of moisture on the hoof. Very wet bedding contains enzymes which break down the cement between the layers of horn. This predisposes to the entry of bacteria and subsequent infections developing in the hoof. These infections are termed white line abscesses. Lack of exercise is also detrimental to the hooves of goats kept standing in very wet conditions for long periods. In this situation the blood supply to the horn is reduced, resulting in poor growth.

Remember that lameness from any cause reduces milk yield, and hence profitability. Owners who experience a high incidence of lameness should contemplate bathing the hooves of their goats in either 10 per cent copper sulphate solution (450 g in 4 litres of water) or 5 per cent formalin solution (200 ml of 40 per cent formalin in 4 litres of water). These solutions have the dual effect of destroying bacteria around the hoof and hardening the horn. Smaller quantities of either substance can be made up and applied to each hoof individually. Avoid splashing these solutions on to the skin or eyes, and do not leave the goats standing in these solutions for longer than two or three minutes.

Summary of Measures to Avoid Lameness in Housed Goats
1. Constant attention to feet paring.
2. Provide as dry a bed as possible with good drainage.
3. If possible provide an area of dry concrete, where the goats can spend part of the day.
4. Bathe feet in either formalin or copper sulphate solution.
5. Exercise daily.
6. Vaccination (*see* Footrot, page 98).

SOME SPECIFIC CAUSES OF LAMENESS

Footrot
Although classically a problem of grazing animals in the summer time, footrot does occur when goats are bedded on straw in winter. If the conditions are warm and wet, the bacterium *Fusiformis necrophorus* can survive to be spread from one goat to the next. (*see* Footrot, page 98).

LAMINITIS

The laminae are the sensitive tissues which lie below the layer of horn which covers the hoof (see plate 3.2). When these tissues are inflamed they become extremely painful, a condition described as laminitis.

Symptoms

Laminitis, like many diseases, can be either very severe or very mild, with all grades in between. Characteristically the goat appears uncomfortable, or severely lame on all its feet, especially the fore feet. The feet can feel hot when touched owing to the inflammation below the horn, and the goat's temperature is elevated. This condition arises in goats receiving heavy concentrate rations and in cases of over-eating (for example when the goat gains access to the concentrates by accident). It is also a complication to many other diseases such as mastitis or an infection of the womb (metritis), especially when these diseases occur close to when the goat kids. Spring grass has also been incriminated as a cause of this disease.

Treatment

Whatever the cause of this condition, treatment from your veterinary surgeon is required. Very severe cases may not respond well to treatment, but with modern drugs it is always worth attempting treatment. Your veterinary surgeon will most probably prescribe drugs which are anti-inflammatory and relieve the pain in the hoof. In instances where the goat is receiving high quantities of concentrate food, the first step to take, even before you ring the vet, is to stop feeding.

Prevention of laminitis

Probably the most important fact to remember is that goats are not really intended to eat concentrate foodstuff. They can do so, but only in relatively limited quantities, and certainly not when it is given suddenly in very large amounts. Remember too, that high-yielding goats, fed maximum quantities of concentrate foodstuff, may in fact be suffering from a very mild form of laminitis. This may pass unnoticed for many months until feet deformities occur and the goats are reluctant to walk. Despite treatment there is a tendency for this condition to recur.

PUNCTURE OF THE SOLE

Dirty objects such as nails which puncture and penetrate the sole of the hoof inoculate bacteria into the laminae. If the bacteria multiply in that site pus is produced which soon leads to pain and lameness. Unfortunately the hard horn of the hoof will not allow the pus out, and the goat continues to be lame. This is another job for your veterinary surgeon, who will open the wound, release the pus and also give injections to destroy any germs that have entered. Another possible result of this problem is, of course, tetanus; so always ensure that your goats are protected against this infection by vaccinating them.

White Line Abscess

As described in the introduction to the section on lameness, constant standing in wet faeces and urine can lead to a deterioration of the 'white line'. This is the junction of the wall and the hoof horn; as shown in fig. 3.3. If bacteria penetrate into this space pus forms, causing pain and lameness. Your vet will search out the pus with a hoof knife, relieve the pain and dress the foot.

Treatment of Lame Goats

Lameness is a symptom of pain somewhere in a limb. It is nature's way of preventing the animal from using that limb and enforcing rest. Although problems can obviously arise anywhere in the limb, in practice most causes of lameness are found in the lower part of the leg, or in the hoof. Examination of the hoof for unusual signs such as swelling, must be the obvious first course of action. Having established whether or not the lameness arises from the hoof region the goatkeeper must decide whether or not the injury is sufficient to warrant attention from a veterinary surgeon. If your goat has any condition which makes you suspect that the problem arises high up in the limb you should refer the animal to a veterinary surgeon. It is impossible to give strict guidance in respect of foot lameness, but the following indications will hopefully be of value.

Suggested Examination and Treatment that can be Given by the Goatkeeper
1. Trim the hoof
2. Wash the hoof, using a stiff brush.
3. Search for stones or other objects between the claws.
4. Examine hoof for penetrating foreign bodies, such as a nail.
5. Look for signs of infection in the soft tissue between the claws and any evidence of a foul smell.
6. Bathe the hoof in copper sulphate or formalin solution; this will rarely do any harm.

7. Rest the goat, giving her special attention at feeding time to ensure that the other goats do not bully her.

Problems Requiring Veterinary Attention

1. Suspected fractures of the bones.
2. Any puncture wound of the hoof, for example where the animal has trodden on a nail.
3. Severe lameness, accompanied by an angry red swelling around the top of the hoof (this indicates a sepsis of the foot).
4. Any lameness that does not respond to conservative treatment given by the owner over two or three days.
5. Any lameness with blisters around the coronet.
6. Goats reluctant to stand, apparently sore on all four feet.
7. Any lame goat refusing food.
8. Splitting of the horn into the coronet.
9. Heat in the hoof.

COCCIDIOSIS

I have included this important problem in this chapter because the condition is most commonly seen in housed goats; indeed housing has a profound effect on its occurrence.

The Coccidia

Agents which are almost always present in the surroundings of goats are the protozoan parasites called coccidia. There are several stages of this organism's life cycle which take place in the cells of the large and small intestines of the goat. When present in small numbers the coccidia cause very little damage to the goat and no disease. Fig. 3.6 illustrates the life cycle of the parasite.

The number of oocysts put out in the droppings of goats is extremely variable. Does can have counts of between 3,000 and 15,000 oocysts per gram of faeces. Kids with or without signs of disease, can have counts as high as 60,000. An immune adult goat could put out 216 million oocysts per day and the potential for infecting other goats is obviously very high. The oocysts in the droppings of one goat become infective to other goats after about one week. On ingestion by other goats, they pass through the stomach and into the intestine. The wall of the oocyst breaks down allowing the parasite to invade the lining of the intestine. It is this damage, destruction of cells and the rupturing of blood vessels which give rise to the symptoms of weight loss and bloody diarrhoea.

Almost all species of animal have their own strain of coccidia and the coccidia of rabbits and chickens, for example, do not cause disease in goats. The coccidia of sheep, however, may be responsi-

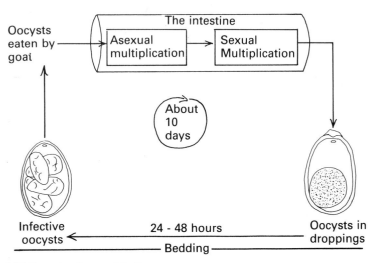

3.6 *Life cycle of a coccidium*

ble for some disease in goats and therefore should be regarded as suspect. The species of coccidia affecting goats include *Eimeria arloingi* and *E. faurei*.

Symptoms

- MILD FORM   Kids off their food with symptoms of diarrhoea.

- ACUTE FORM   Sick goats with blood in diarrhoea. The animals are dehydrated and they may show persistent straining in their attempts to pass faeces. Your veterinary surgeon may send samples of the goats' dropping to a laboratory in order to confirm his diagnosis. Goats that recover from coccidiosis may become unthrifty.

- VERY ACUTE FORM   Recent experimental work suggests that a very acute syndrome can occur resulting in death of the kid within twenty-four hours. Workers found numerous nodules in the small intestines at post-mortem examination. These were sites of haemorrhagic enteritis. No digestive symptoms were observed. The French workers propose that such nodules may become activated by various stress factors and favour the development of enterotoxaemia.

Treatment
Response to drugs given by mouth can be good but treatment must be initiated quickly. Commonly used drugs include sulphonamides and Amprolium. Sulphonamides are probably the drug that your veterinary surgeon would prescribe because they have the dual function of controlling the coccidia and at the same time preventing secondary bacterial infection.

When large numbers of goats are being treated, sulphonamides can be mixed in with the feed or drinking water at a dose rate of 145mg/kg body weight daily. Amprolium used similarly requires 25 to 50mg/kg of body weight daily.

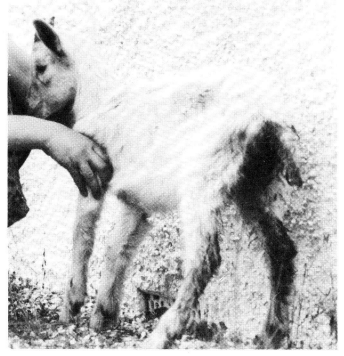

PLATE 3.5
Coccidiosis, complicated by worms

Prevention and control
It is possible to avoid this disease and the answer lies in good management. As a rule of thumb, regard all adults as infected and immune, and all kids as extremely susceptible. The aim of good husbandry is to allow the kids gradual exposure to the agent. This is achieved naturally out at pasture so that the disease is rare when goats are grazing large areas.

Disease is generally seen in housed kids. It can occur in circumstances where they are exposed to infection at a late stage. Gradual disease-free exposure normally occurs in kids housed in clean pens. The kids experience a small challenge but immunity develops quickly and no disease is seen. Reinfection occurs after ten days and still the kids' developing immunity keeps up with the challenge.

There are three situations when disease frequently occurs.

1. When the bedding is very wet and the house is warm and humid. These conditions favour the parasite and allow build-up of oocysts which can overcome the kids' defences.

2. Where pens are not cleaned out between batches of kids. In this instance the first batch of kids grow up with the increasing levels of infection but their increasing immunity matches it. The second batch of young (susceptible) kids soon falls prey to the high levels of challenge, and disease results.

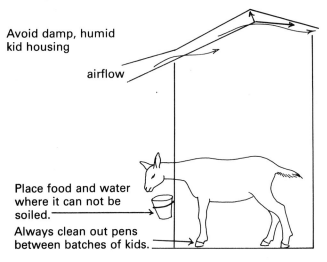

Avoid damp, humid kid housing

airflow

Place food and water where it can not be soiled.

Always clean out pens between batches of kids.

3.7 The prevention of coccidiosis

3. Where food and water containers become contaminated. Poor design in the siting of food and water containers means that droppings contaminate food and water so that the kids are forced to take in large numbers of oocysts which can result in disease.

The Use of Drugs to Prevent Coccidiosis

Some authors have recommended the use of drugs such as Amprolium to control coccidiosis in groups of kids, where the problem keeps recurring. Medication is given for several days every three weeks, the objective being to assist the kids face a heavy challenge by eliminating the infection while allowing immunity to build up. In my opinion the indications for this are extremely rare, an example being when young goats have to be mixed with older ones. In most cases sensible changes in husbandry would eliminate the need for such medication.

Worm Control in Yarded and Stall-Fed Goats

Parasites are rarely a problem in adult yarded goats, but it depends upon what is fed to them. The reader should refer to page 88.

Chapter 4

PROBLEMS ASSOCIATED WITH FEEDING

No chapter on this topic would be complete without first of all giving the reader a basic understanding of the anatomy and functioning of the goat's digestive system. The goat belongs to the same family as sheep and cattle, all the members of which are described as 'ruminants'. They are so called because they possess a complex digestive system which incorporates a large stomach or rumen. The function of this organ will be outlined in the text below. Ruminants are therefore very different to simple-stomached (monogastric) animals such as the dog or pig.

THE ANATOMY OF THE GOAT'S DIGESTIVE SYSTEM

For the sake of clarity this description will commence at the mouth and progress towards the back passage in sequence.

Examination of the mouth of the goat is difficult but the reader should be able to note that unlike humans the goat lacks any incisors or front teeth on the upper jaw. The goat grazes or browses by biting with the front teeth of the lower jaw against a hard pad on the top jaw.

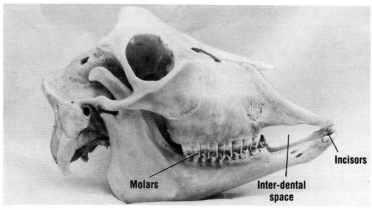

PLATE 4.1
The dentition of a goat, showing the absence of the upper incisors

Assisted by the tongue, the food is then passed back to the cheek teeth where it is given a brief chew before being passed into the oesophagus and on to the stomach The oesophagus is simply a muscular tube connecting the mouth to the stomach. The stomach is divided into four compartments; listed in the order in which food passes through them they are: 1. rumen, 2. reticulum, 3. omasum and 4. abomasum. Plate 4.2 illustrates their relative size.

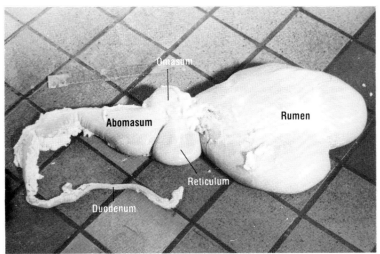

PLATE 4.2
Stomachs and duodenum of an adult goat (compare plate 2.1)

Such a digestive apparatus enables ruminants to utilise grass, leaves and other vegetable matter which simple-stomached animals such as a dog are unable to use. After a period of grazing or browsing the goat commences to ruminate or 'chew the cud'. This involves moving food material back up the oesophagus to the mouth where it is chewed between the goat's back teeth. Following this chewing, the food is returned to the rumen where it remains for several hours undergoing fermentation. Gas is produced as a result of this fermentation and this must be eliminated by the goat belching frequently.

It is during this fermentation process that nutrients are unlocked from the vegetable matter of the goat's diet. The contents of the cells of plants are not available to most monogastric animals because they lack the enzymes necessary to digest the plant cell wall. Goats also lack these enzymes but they solve the problem by allowing bacteria and micro-organisms to do the job for them.

Goats provide the bacteria with a suitable warm place to live, the rumen, and in return the bacteria release, for the goat's benefit, the food locked up in the plant cells. After the food has been fermented in the rumen and reticulum it gradually passes through the omasum or 'bible', where further grinding of the food particles takes place.

From here it passes into the abomasum or true stomach, an organ not dissimilar to our own. Digestive enzymes mix with the food in the abomasum and this muscular sack pushes food through into the small intestine at regular intervals.

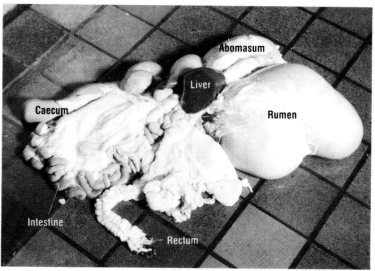

PLATE 4.3
Stomachs and intestines of an adult goat

In the small intestine further digestive juices are secreted along with those from the liver and pancreas. A great deal of absorption of nutrients occurs in this part of the digestive tract.

The caecum is a blind-ending sack in which some absorption of fluids takes place. The large intestine functions mainly to resorb water from the products of digestion which pass through it. The familiar 'pellets' of the goat form in the last part of the large bowel or colon.

There are several important comments one can make at this stage in relation to this digestive system.

1. The goat must always be able to get rid of the gas otherwise bloat develops (*see* BLOAT, page 109).

2. The rumen is a large fermentation chamber, full of many beneficial micro-organisms which live in a delicate balance.
3. The capacity of the rumen is very great: two gallons in the adult goat. Should a goat gain access to concentrate foodstuffs by mistake she can consume vast quantities, more than is good for her.
4. Adult goats vomit with difficulty, and therefore anything taken into the rumen by mistake is well and truly locked within the animal.

Kids
The diet of kids is of course milk, and the complex stomach of the adult animal is not required. Up to about six weeks of age only the abomasum or true stomach is working and the kid's digestive system functions rather like that of the monogastric animal (*see* Chapter 3 'Problems Associated with Kids').

TEETH AND PROBLEMS OF SORE MOUTHS

Goats may refuse food for two basic reasons. Firstly they may feel ill and disinclined to eat and secondly they may find the eating process painful. Typical reasons for the latter are painful teeth or injuries to the soft tissues of the mouth. Goats with mild pain in the mouth may just eat very slowly or drop food from the mouth. They may drool saliva and 'smack' their lips. In such instances the goatkeeper should examine the mouth in order to try and determine the cause.

Unfortunately it is very difficult to open the mouths of goats in order to make such an examination, and it may be necessary to sedate the goat. Remember not to place your fingers between the back teeth as this can result in a painful injury. Swellings of the soft tissue of the mouth frequently result from small injuries which sometimes become serious enough to warrant an injection of antibiotic.

'CHOKE', OESOPHAGEAL OBSTRUCTION

Occasionally goats suffer from choke when an object of food material obstructs the oesophagus, preventing belching. This is a potentially dangerous situation because the build-up of gas in the rumen can very quickly cause death (*see* BLOAT).

Action to be taken in the event of choke
Call your vet who will probably try to relax the oesophagus using a smooth muscle-relaxant; he may pass a stomach tube in order to

dislodge the obstruction. If this action fails he may be obliged to perform a small operation in order to remove the offending object. This is done by cutting into the oesophagus from the outside. *See also* page 179.

Bloat
see Chapter 5, 'Problems Associated with Grazing', page 109.

INDIGESTION

Occasionally goats in peak lactation may suffer from indigestion resulting in loss of appetite. There is failure of the normal rumenal movement and cudding ceases. It is usually associated with high intake of concentrate foodstuffs. Many mild diseases may cause indigestion but are of little significance.

Symptoms
The goat is off its food, slightly dull and suffers a small fall in milk yield. There are few other signs, and generally the goat recovers within two days.

Treatment
Generally this is not essential but mild rumenal stimulants may help. Sodium bicarbonate given by mouth may be of some use if there is a tendency for acid conditions in the rumen.

ACIDOSIS

This is the term given to the condition when the rumen contents become acid. It generally occurs after a goat has accidentally taken in large quantities of concentrate foodstuffs when, for example, the store shed has been left open by mistake and the goatkeeper finds the goats eating buckets of dairy cubes. This is probably one of the most serious accidents that can befall a goat. The rumen becomes full of concentrates which quickly ferment to acid.

An important factor in this disease is the previous diet of the goat involved. A heavy milker that is receiving several kilograms of concentrate may be relatively unaffected by this quantity, because her rumen is adapted to it. On the other hand, a goat at pasture which suddenly consumes that quantity of feed may become extremely ill.

Symptoms
These can vary depending upon the previous diet of the goat and on the quantity of food consumed. In the early stages the goat becomes

depressed and hangs its head. Later the goat becomes intoxicated and 'drunken' in behaviour. Bloat tends to occur and there is a swelling on the left flank (*see* page 109). Because of the pain the goat may grind its teeth and diarrhoea soon becomes apparent. Eventually the goat 'goes down' and is unable to rise, a truly sorry state. Very acute cases die within twenty-four hours.

Treatment
This condition must be treated as an emergency by the goatkeeper. Veterinary attention must be sought if the animal is to have a chance of survival. The first thing to do is of course to stop access to any more food. A first-aid measure would be to drench the goat with something alkaline such as bicarbonate of soda. Two or three ounces will help neutralise the acid. Walking the goat up and down is probably of some value as well.

Your veterinary surgeon may decide upon a number of different treatments, depending upon the case. He may choose to operate on the goat in order to remove the foodstuff directly from the rumen. He will probably only do this if the goat is still in reasonable shape because dehydration occurs, making the animal a poor surgical risk. Many cases are treated by drugs alone. Fluids obviously help in dehydrated cases and vitamins help to detoxify the animal. Sometimes antihistamine drugs are given to overcome laminitis.

DISPLACEMENT OF THE ABOMASUM

Very rarely, a condition occurs when the fourth stomach moves from the right-hand side of the abdomen to the left-hand side. It becomes trapped between the rumen and the left flank wall, giving rise to a specific syndrome.

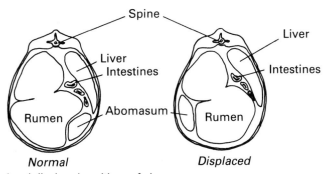

4.1 Normal and displaced positions of abomasum

The Reason for Displacement
The condition is thought to occur just after kidding. During pregnancy the uterus expands, pushing the abomasum under the rumen. The rumen also tends to rise as a result of this pressure. It is believed that after kidding the rumen falls down, trapping the abomasum.

Symptoms
The condition would normally occur around parturition. The appetite of affected goats becomes reduced; they tend to pick at their food and refuse concentrates. Milk yield would obviously be reduced. Your veterinary surgeon will confirm his suspicion by listening to characteristic sounds on the left flank.

Treatment
In cattle, where the disease is common, two forms of treatment are used; one involves surgery, the other does not.

1. *Surgical repair.* The surgeon pulls back the fourth stomach and stitches it to the right-hand flank wall.
2. *Rolling.* In cattle, rolling the animal on its back in the correct manner can restore the stomach to its former position. I have a suspicion that some cases resolve themselves during the journey by car to the vet's surgery. On more than one occasion symptoms of this condition have been described to me on the telephone but the goat appears 'cured' when she arrives!

Prevention
There are precious few records of this condition in goats so one must look to work in cattle for some guidance. One may postulate that avoiding heavy concentrate feeding during pregnancy may help to avoid the problem occurring. The main reason for this is that heavy concentrate feed will result in a small rumen (less bulk) and therefore allow the displacement to occur more easily.

AFLATOXICOSIS (MOULD POISONING)

Foodstuffs such as groundnuts, soya beans and cereals sometimes become contaminated by fungus or mould. The toxins damage the liver, reducing its efficiency. Symptoms include loss of condition, reduced appetite and possibly jaundice. In an incident in Sri Lanka, 194 out of 1800 kids died. They were mainly aged between 6 and 9 months, and at post-mortem examination the livers were found to be destroyed, being hard and fibrous.

PREGNANCY TOXAEMIA

This is a metabolic disease of does in late pregnancy. It is non-infectious, being a product of disturbed carbohydrate usage. It can be caused by either faulty feeding or starvation.

Figure 4.2 shows the normal situation for a non-pregnant, non-lactating doe. She represents an animal with the least demand for nutrients because all she requires is enough to maintain her tissues in working order.

4.2 The energy requirements (maintenance) of a non-producing doe

A goat in late pregnancy obviously has extra demand upon the supply of food especially for carbohydrate (energy) foods, as shown in fig. 4.3.

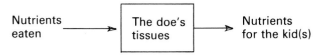

4.3 The energy requirements (maintenance and production) of a pregnant doe

If in fact the doe is carrying twins then the burden may be too high unless the food contains a large amount of energy. Some foods — for example, poor-quality hay — may have enough energy for a non-pregnant goat, but insufficient amounts for a doe carrying twins. Also in late pregnancy the uterus and its contents take up a large amount of space in the doe's abdomen so that she cannot possibly eat enough poor-quality foodstuff to provide all her requirements.

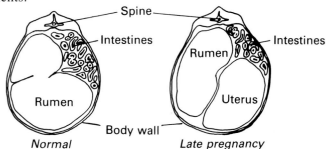

4.4 Effect on the rumen of the uterus of a heavily pregnant doe

When this state of affairs occurs the doe ensures that the kids get enough carbohydrates at the expense of her own tissues. She then tries to rob her own reserves of carbohydrates and this leads to the release of ketone bodies into her blood. The ketones are a sign that her metabolism is faulty.

Symptoms

The symptoms of pregnancy toxaemia are those of blindness, reluctance to move and loss of fear. Affected does go off their feed and bump into objects when moved. Later they 'go down' and may even suffer from convulsions. If untreated they will die within a week.

Treatment

Response to treatment is not very good, although it is well worth attempting it. There are two basic objectives: 1. to give the doe a readily usable from of energy such as glucose; 2. to get her eating again (using anabolic steriods such as trenbolone acetate).

Of course she still has twins inside draining away any energy you give her and so many vets either encourage the doe to deliver her kids early (alive or dead), or perform a Caesarian section. The decision of which to do depends upon the value of the kids, the stage of the pregnancy and the condition of the doe.

As a first-aid treatment, before the vet arrives, the owner could drench the goat with glucose in water or molasses (treacle) in water. 100 grams of either would do some good.

Prevention

1. Feed good-quality forage to does in the last two months of pregnancy.
2. Always supplement their forage in the last eight weeks of pregnancy with some concentrate (in case they are carrying twins or triplets). Remember that three-quarters of goat pregnancies result in the birth of two or more kids.

   The nutrient requirements for a pregnant goat weighing 70 kg two months before kidding are:

Drymatter kg/day	ME MJ/day	DCP g/day
1.5	15.5	121

3. Daily exercise may be useful.

ACETONAEMIA (KETOSIS)

This condition is remarkably similar to pregnancy toxaemia described above. The main difference is the time when acetonaemia manifests itself, being characteristically in the first month of lactation. Another feature is that it is a disease of housed does in the winter months.

Why the Disease Occurs

Referring back to figs. 4.2 and 4.3, the lactating doe is similar to the pregnant doe in terms of requirement for carbohydrate because a great deal is required for milk production. This could be represented as fig. 4.5.

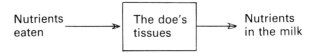

4.5 Energy requirements of a lactating doe

Obviously the high-yielding doe in peak lactation (about 2-3 weeks after kidding), requires enormous amounts of energy feed. If she is unable to obtain this, the same circumstances prevail as in pregnancy toxaemia. Ketones (acetones) accumulate in the blood and this is described as acetonaemia.

Symptoms

The doe goes off her food, tending to prefer hay to concentrates. Milk yield falls and the goatkeeper may be able to detect a sweet smell in the goat's breath (this is the smell of the acetone).

Treatment

This is more simple than treating pregnancy toxaemia and the condition can respond to treatment very quickly. On the other hand many stubborn protracted recoveries have been widely reported. Again, as a first-aid treatment, a glucose or treacle drench may be given. Your veterinary surgeon may inject the goat with a corticosteroid drug which usually dramatically reverses the condition. He may prescribe easily used, high-energy substances, such as propylene glycol to be given by mouth. Multivitamin injections may also be of value. Once the goat has regained her appetite increase her ration of energy foods (cereals) so that relapse does not occur.

Prevention
This is mainly aimed at ensuring adequate levels of concentrates for high-yielders. It may be necessary to feed concentrate little and often just after kidding because the doe may be finding it difficult to eat enough food to keep up with the tremendous ouput of milk. You can help by giving the goat daily exercise and possibly by feeding titbits of green foods to stimulate appetite. You may learn a great deal about avoiding this condition by discussing feeding with a local dairy farmer. The principles are the same for dairy cows and goats.

Dietary Scour (Diarrhoea) in Adult Goats

Occasionally, changes in diet result in diarrhoea in the adult goat. Every animal has a gut population of micro-organisms which in normal circumstances live in a stable situation. Following a change in the type of food certain bacteria begin to multiply and take over, because the conditions become just right for them. Such a situation can result in toxins irritating the gut lining, causing massive out-pourings of fluids. At the same time absorption of food declines and diarrhoea results.

The Danger of Enterotoxaemia
Such circumstances are ideal for the development of enterotox-aemia. The conditions allow the multiplication of certain clostridial bacteria. Remember to have all goats vaccinated against this disease.

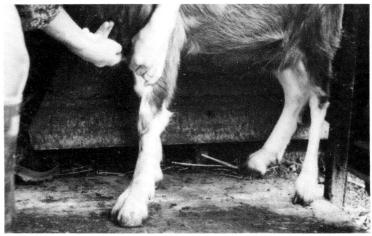

PLATE 4.4
Clostridial vaccination on the inner surface of the foreleg

Symptoms of dietary diarrhoea
Generally, an uncomplicated dietary scour gives rise to few other symptoms, although there may be a fall in milk yield in lactating animals.

Treatment
Cutting food back is important, especially restricting the quantity of the new type of food material. In many cases this action alone will restore the goat to normality. Hay is a very useful feed for goats suffering from diarrhoea; oak leaves help, but have a constipatory action. In more severe cases all sorts of remedies are available including kaolin. Antibiotics should not be necessary but your vet may give the goat an injection of drugs such as Buscopan (Buscopan Compositum Hyoscine.N. butylbromide plus dipyrone) which slows down the passage of fluids through the intestine.

Prevention
It seems very obvious but it never does any harm to stress the need to introduce new dietary compounds slowly. This allows the micro-organisms in the gut to adapt gradually to the new food.

INTESTINAL OBSTRUCTION (BLOCKAGE)

This is a fortunately rare event in the goat which I have never seen, but I shall describe for the sake of completeness. It could result from many causes such as a telescoping effect of the bowel (called an intussusception) or, possibly, adhesions. One case was caused by a plastic bag. The symptoms would be those of a sick goat which has no appetite, possibly a distended abdomen and no droppings being passed. Because of the pain, teeth-grinding and belly-kicking might be noted.The early signs may be restricted to paleness of the membranes and a sub-normal temperature.

CEREBRO CORTICAL NECROSIS (C.C.N.)

At first sight it may seem strange to include a condition of brain malfunction in this chapter on feeding but the two are related. This condition gives rise to nervous symptoms in goats of any age. The symptoms result from necrosis or death of some of the cells in the cortex of the brain. The explanation for the death of these cells is a lack of thiamine (vitamin B1).

Thiamine is normally present in adequate quantities in the rumen. In certain circumstances, however, fungi present in the food destroy the thiamine with their enzymes. Thus goats fed on foods containing large numbers of fungi can become deficient in thiamine and hence develop symptoms.

Symptoms
The goat shows nervous symptoms such as blindness and also standing with the head pressed against a wall. They may wander in circles and even develop convulsions. These symptoms are similar to those of other diseases such as lead poisoning so your veterinary surgeon will probably send samples to a laboratory for diagnosis.

Treatment
The administration of thiamine is obviously useful but the response to treatment may be poor if the brain is too badly damaged.

One report from the United States declared that only three out of six treated goats responded to thiamine therapy. The other three died between two and seven days after the onset of symptoms. All the goats were being fed concentrates and the outbreak occurred during the winter.

Listeriosis
Goats fed on silage are prone to develop symptoms of listeriosis, an infection by *Listeria monocytogenes*. The conditions produced when making silage are often conducive to the proliferation of the organism. (*See* LISTERIOSIS, page 164.)

DEFICIENCIES OF ESSENTIAL DIETARY COMPONENTS (VITAMINS AND MINERALS)

VITAMINS

The vitamins are a group of compounds that have been shown to be essential for the normal functioning of the goat and other animals. They are only required in very small quantities but if they are absent from the diet the effect can be quite dramatic. Unlike monogastric animals goats are able to manufacture some of their vitamins, or at least the rumen micro-organisms do it for them. Vitamin D can of course be produced by the skin in the presence of sunlight. The stress laid upon vitamin deficiencies by the manufacturers of vitamin products is probably over-emphasised. That does not mean to say that goatkeepers should not be aware of possible deficiency states.

The Diet and Vitamin Supply

Vitamins are present in varying quantities in different food substances. It is fairly obvious, therefore, that a goat fed a varied diet is more likely to avoid the pitfalls of vitamain deficiency than is a goat whose diet is more restricted. By feeding a little of this substance and a little of that, deficiencies can normally be avoided, even if the goatkeeper is unaware of the specific components of each kind of foodstuff.

Stress, Disease Conditions and Vitamins

Bacterial infections increase the demand for some vitamins such as vitamin A. When animals are ill they are generally off their food and hence, in sickness, goats may benefit from extra vitamins. Stress conditions are also thought to increase the requirement for vitamins.

VITAMIN A

This vitamin is essential for the health of all the lining surfaces of the goat's body; for example, the skin and the cells lining the gut. It is taken in by the grazing goat in the form of protovitamin A and is converted into vitamin A mainly by the intestinal wall. Good pasture contains adequate quantities (between 3 and 8 mg per cent) and fresh hay between 1 and 2 mg per cent. Reserves can be stored in the liver for a few months. Calves born to cows grazed on pasture have higher levels than calves born to stable-kept cows. This is probably true of goats and their kids.

Symptoms of deficiency
Because this vitamin is essential for the lining surfaces, a lack of it can cause deficiencies of any of these surfaces (the gut, windpipe and bronchi, urinary tract etc.). Thus diarrhoea and respiratory disease will be likely. Blindness may also occur. In one recorded problem in a herd of 223 goats the following symptoms predominated: abortion, diarrhoea and the loss of sight due to the opacity of the cornea. The problem is likly to occur only on very poor pasture or when yarded goats are fed very poor diets for long periods.

Treatment
Response to treatment with vitamin A is generally good. Toxicity resulting from too much vitamin A has been occasionally recorded but it is most unlikely to occur unless doses of 200 times the normal level are given over long periods. A suitable time for dosing is late pregnancy, which will benefit both the dam and the kid.

VITAMIN D

This vitamin regulates the calcium and phosphorus in the animal's body. It acts in several ways, which include increasing the mineralisation of bones and teeth. It is present in green foods but low in stored hay. Animals can manufacture their own vitamin D in the skin, in the presence of sunlight.

Symptoms
Poor growth and feed conversion. The disease is termed rickets in the young and osteomalacia, or bone softening, in the adult. Osteomalacia is very rare but gives symptoms of stiffness and crippling.

RICKETS

Rickets can occur in young rapidly growing goat kids, although it is uncommon in the United Kingdom. It can result from a deficiency of either calcium or phosphorus of vitamin D. When any one of these factors are missing from the diet defective bone formation occurs.

Symptoms
Stiffness of movement characterises this condition, accompanied by enlarged joints, especially of the front legs. It may also be possible to feel a swelling half way down each rib, forming a chain along the chest. The joint swelling results from poorly mineralised bone undergoing pressure and becoming deformed.

Treatment
Providing a balanced diet containing calcium and phosphorus is the key to success with treatment. Exposing the animals to sunlight enables them to manufacture their own vitamin D in the skin. Injections of vitamin D may be given by your vet.

Prevention
As stressed under *Treatment*, a well-balanced diet, containing minerals, is essential. If one is ever concerned about possible deficiencies in feedstuffs remember that by ensuring variety in the diet one often avoids deficiency.

VITAMIN E

Vitamin E is present in hay and grass and to a lesser degree in cereals. Deficiency is only likely to occur in animals kept on poor-quality rations. The activities of this vitamin are poorly understood, though it is considered to act as an antioxidant. Its functions are closely bound up with the mineral selenium. In some circumstances either selenium or vitamin E will prevent problems. Soils and pastures vary widely in their content of selenium.

Symptoms of deficiency
A deficiency of this vitamin disturbs both cell metabolism and the structure of all the component living parts of the cell. The principal symptoms of deficiency are associated with muscles, that is, the degeneration of heart and skeletal muscle. Stress and disease conditions are thought to increase the requirements for this vitamin.

Two types of symptom are described, mild and acute. In the mild form stiffness, weakness and trembling are reported. The acute form in the goat is described where the affected animals are found dead without showing any previous signs. Thus, most of our knowledge of the disease comes from work on dead animals, at post-mortem examinations. When examined, all the evidence of muscle damage can be seen in the heart, diaphragm and legs. White streaks of unhealthy tissue can be seen among the normal muscle fibres. Unfortunately tests for the deficiency are not useful on a routine basis and most goatkeepers only learn of problems after deaths have occurred. Emphasis must therefore be placed on prevention of the disease.

Treatment
Normally the administration of selenium, together with vitamin E, will reverse the symptoms within a week. Vitamin E preparations can be given either by mouth or injection.

Prevention
A varied diet with a plentiful supply of good-quality hay, made while the plant is still green, will normally ensure freedom from the disease. The condition is extremely unlikely to occur while animals are at pasture. Yarded animals, fed on diets such as roots known to be low in vitamin E, should receive supplementation in their diet.

Vitamins C and K

Both these vitamins are present in grass and hay. They can also be manufactured by the bacteria in the rumen (for vitamin K) or in the liver and kidneys (for vitamin C). Deficiencies of these vitamins, therefore, do not occur and no extra supplements need be given.

MINERALS AND DIET

Minerals such as calcium and phosphorus are obviously essential as components of bone and teeth. They are required in large amounts, especially in the growing kid, which is rapidly building up bone tissue.

Trace Elements

Some minerals are only required in minute quantities and, as only a trace is needed for healthy functioning, they are described as 'trace elements'. Manganese and copper would be examples of these elements.

The Turnover of Minerals

The complement of minerals within the goat's body is never static but is continuously changing. Minerals are continually being taken in along with the food and simultaneously lost from the body. In order to remain healthy, however, the delicate balance must be maintained and very complex regulation takes place. This regulation is sometimes carried out by hormones such as the parathyroid hormone. The important fact to note is that minerals must always be available for use by the goat. It is no use giving large quantities at intervals and then depriving the goat for weeks.

Mineral Deficiencies

In well-managed goats, fed a variety of foodstuffs, deficiencies are very unusual. Well-documented laboratory experiments have been conducted so that scientists do know the effects of deficiency of numerous minerals but in the field these are extremely rare. Feeding imbalances occur from time to time and some of them are dealt with in the following section. In such instances, errors often arise as a result of feeding one nutrient in quantities which influence another nutrient.

COBALT

Cobalt is essential for the production of vitamin B12 by micro-organisms in the rumen (*see* page 64). Deficiency leads to a rather vague wasting condition in sheep and goats called 'pine'. The lack of vitamin B12 (cyanocobalamin) leads to loss of appetite, and thus affected goats simply do not eat and they waste away.

Symptoms

Loss of body weight, poor appetite and anaemia are classic

symptoms. This condition may be compounded by worms, because the deficiency makes them more susceptible to roundworm parasites. The goats can become so weak that they literally pine away and die.

Treatment
Cobalt can be given by mouth or vitamin B12 may be given by injection. 'Bullets' containing cobalt are available but they are only of use in kids over two to three months of age. These allow the slow release of cobalt into the the rumen contents.

Prevention
The use of cobalt bullets is the most convenient method of preventing this disease. Areas of the United Kingdom affected by cobalt deficiency are well known to veterinary surgeons so your vet will probably give you guidance on this aspect.
(*see also* PINE, page 108.)

COPPER DEFICIENCY

The effects of copper deficiency on pregnant does result in their giving birth to kids with swayback (*see* ENZOOTIC ATAXIA, page 39).

In cattle and sheep, deficiency of copper also gives rise to a variety of symptoms such as anaemia, poor growth and loss of milk production. Although I can find no documented reports of such conditions in goats, other authors do suggest that they may exist. Verbal communcations support this opinion.

It may be important to note that another mineral, molybdenum, has an action directly opposite to that of copper and therefore copper deficiency has been recorded in areas where molybdenum levels are high.

Treatment and prevention
If copper deficiency is confirmed goats can be given copper either by injection or by dosing by mouth. Copper is very toxic, however, and guidance should be obtained from your veterinary surgeon on how much to give.

Poisoning by Copper
The metabolism of copper in the goat appears to be different from that in sheep. Goats are in fact more resistant to copper poisoning than sheep because less copper accumulates in the liver.

IODINE DEFICIENCY

Occasionally this element can be deficient in the diet, giving rise to 'goitre' or swollen thyroid glands. It may also be induced because the animals are consuming goitrogenic foods such as kale. Goitrogenic substances lock up the iodine in the food and cause a deficiency. Iodine is essential for the production of thyroid hormone, an important body regulator which has a profound effect upon the rate of the body's chemical reactions.

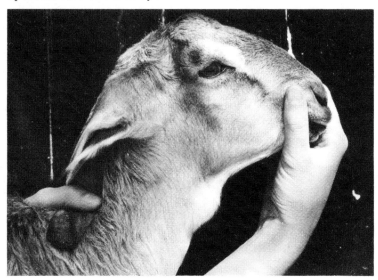

PLATE 4.5
Thyroid enlargement produced by iodine deficiency

Symptoms
They may be most obvious in the newborn kids which show weaknesss, absence of hair and large thyroid glands. It may be seen in fast-growing kids because milk is a poor source of iodine. The thyroid glands may also be enlarged in the adult goats.

Treatment and prevention
This condition is reversible and it is always worth attempting to treat affected animals. Because of the toxicity of iodine, treatment doses should only be as large as doses aimed at prevention. Feeding iodised salt (containing at least 0.007 per cent iodine) is one method of prevention. Alternatively potassium iodide can be given by mouth or tincture of iodine can be applied to the animal's coat once

a week. This will be licked off by the goat. For oral administration 2ml of a 2 per cent solution of iodine in potassium iodide gives good results.

SALT (SODIUM CHLORIDE) DEFICIENCY

Sodium and chlorine are essential for maintenance of the tissue fluids and they are vital for numerous bodily conditions such as nerve transmission and water balance. I know of no specific deficiency reports in goats but in heavily lactating dairy cows, deficiency is thought to be widespread. Proprietary concentrate rations, such as dairy cubes, contain 1 per cent sodium chloride.

Prevention
Give goats access to salt in the form of a lick or loose crystals, expecially when they are lactating and at pasture.

ZINC DEFICIENCY

This condition has been produced experimentally in goats; the symptoms were those of a thickened skin. Dietary supplementation would reverse the condition should it occur naturally.

SELENIUM DEFICIENCY

The effects of deficiency of selenium are to produce muscular dystrophy. There is an intimate relationship between vitamin E and selenium; either substance will reverse a deficiency state of either vitamin E or selenium (*see* VITAMIN E, page 76.)

Selenium Toxicity
A toxicity from excess selenium can occur from eating herbage with high levels of selenium but this is very rare and has only been reported in Ireland. In the acute form, nervous symptoms are noted and in the chronic form, lameness and weight loss are reported.

DISEASES ASSOCIATED WITH DEFICIENCY AND IMBALANCE OF PHOSPHORUS, CALCIUM AND VITAMIN D

A disturbance of intake or metabolism of either of these two minerals tends to be very complex. There is also the interrelationship between them and vitamin D. Deficiencies of vitamin D have been covered on page 76. Sometimes enough calcium and phosphorus are fed in adequate quantities but they are fed in the wrong proportions. The suggested calcium to phosphorus ratio is between 2:1 and 1:1 calcium to phosphorus. Large amounts of calcium are present in the teeth and bones. Disturbances show up in the teeth and bones when they are fed incorrectly.

Sources of Calcium and Phosphorus

Calcium is normally present in the leafy parts of plants where phosphorus is relatively low. Phosphorus is found in relatively high levels in grains. Calcium is normally added to the proprietary concentrate rations sold for feeding to lactating dairy animals.

OSTEODYSTROPHIA FIBROSA

This condition is caused by feeding too high a ratio of phosphorus to calcium. It results in a decalcification of the bones, particularly of the skull. Growing animals are most seriously affected. It is rare in goats but may occur as a result of faulty feeding.

Symptoms
Progressive swelling of facial bones occurs (*see* plate 4.6), the bones are soft and 'rubbery'. It has been seen in goats fed on high cereal diets with little green food.

Treatment
Correct the diet to restore the normal ratio of 2:1 calcium to phosphorus.

PLATE 4.6
Osteodystrophia fibrosa, showing the swollen cheeks

DEGENERATIVE JOINT DISEASE

In cattle, especially housed bulls, an arthritis occurs as a result of the excessive feeding of calcium. It is also described in goats, especially bucks, which are never subjected to the heavy calcium losses of pregnancy and lactation. The symptoms are stiffness and a reluctance to move, becoming progressively more severe with age. To avoid the problem care should be taken not to over-feed calcium to housed bucks. Dairy rations are of course formulated for lactating animals and adult males do not require such large quantities of calcium.

BENT LEGS IN GOATS

In Australia a condition described as 'bent leg' developed in young pedigree bucklings aged four months. The forelegs bent inwards or outwards but the animals were in good condition. Experiments suggested that a calcium/phosphorus imbalance was the cause.

IRON

It is extremely unlikely that goats would ever be deficient in iron unless as a complication of the effects of blood-sucking worms. Trials have been carried out on kids to investigate a possible response to extra iron but it was shown to be of no benefit.

Chapter 5

PROBLEMS ASSOCIATED WITH GRAZING

INTERNAL PARASITES

ROUNDWORMS (PARASITIC GASTROENTERITIS)

Roundworms that infest the stomach and intestines can be a serious problem in goats. They cause damage by sucking blood or reducing the absorption of digested food materials from the gut. Reduced appetite has also been shown to result when they are present in large numbers. The deleterious effects, of course, lead to reduced milk yield and smaller weight gain. Inapparent or sub-clinical infestations are thought to be common, which means that the goat does not perform quite as well as she would have done without the worm burden. The common observation that milking does often respond to anthelmintic treatment is borne out by research. In one trial a 17.6 per cent increase in milk yield was observed over a three-week period.

The Physical Appearance of Roundworms
There are various families of worms that infest goats, including strogyles, trichostrongyles and nematodirus. They vary in length from 5 to 30 millimetres. some being visible to the naked eye.

The Challenge to Goats
Goats acquire the infective larvae of these roundworms from the herbage of the pasture as they graze (*see* Fig. 5.1). The larvae pass into the stomach or intestines and remain there, quickly maturing into adult worms. The adult worms then feed (causing damage to the gut) and eventually reproduce and lay eggs. The eggs pass out in the droppings of the goat on to the pasture. The eggs hatch and produce larvae which eventually become infective to other goats.

Disease caused by worms only results when the animal takes in large numbers of infective larvae at one time. A 'trickle' of infection does no harm; in fact it is probably beneficial to the goat, because it stimulates immunity.

The temperature and moisture of the weather influence the development of larvae on the pasture. Hot, dry weather kills larvae; warm, moist conditions help their survival. This means that the weather dictates when the larvae are ready to infest goats. Weather conditions are general for the region and therefore influence the development of all larvae. If the conditions are just right, all the

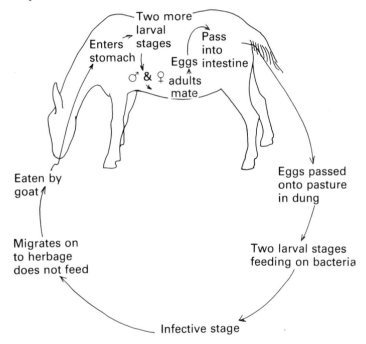

5.1 Life cycle of the gut roundworm in the goat

larvae are infective at the same time and impose a heavy challenge on the grazing goat. Because weather conditions from year to year follow a similar pattern we can predict, fairly confidently, when pastures are dangerous. Thus we can list certain danger periods and suggest ways of avoiding disease. Probably the most important discovery in this field this century has been that there is only one generation of worms each year in the United Kingdom. A further complication, however, is that there is more than one 'group' of larvae each year, and consequently there are several danger periods.

The Sources of Challenge

1. Larvae overwinter which results in a generation of infective larvae on the pasture in late May and June.
2. The infective larvae that result from the post-parturient (after kidding) rise in egg production by the does. This results in a peak of challenge mid July. (The post-parturient rise is a phenomenon seen in goats and sheep when the output of parasite eggs increases dramatically. Various mechanisms account for its occurrence; the parasites lay more eggs and the goat appears to become less resistant to infection at this time.)
3. The larvae passed out over the summer resulting in a challenge in early September.

The danger periods are shown in fig. 5.2. Goats should be removed from dangerous pasture at these times and put onto clean pasture. See later suggested worm control strategies.

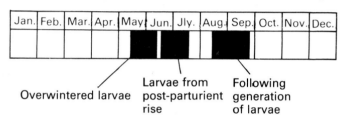

5.2 *The main danger periods for roundworm infestations*

NEMATODIRIASIS

The nematodirus worm is present in a small percentage of goats in the United Kingdom. This worm parasite behaves slightly differently from those in the other group, infestation passing from the kids of one year to the kids of the following year. Thus, if the pasture is used for young kids every year, problems may arise. In fact, in the United Kingdom most kids are reared indoors and only go out to pasture when the threat of disease has disappeared. Goatkeepers in either of the following categories may have kids at risk.

1. Those who graze kids from birth on the same pasture every year.
2. Those who kid very early in the year and turn kids out to the same pasture in April or May each year.

In these instances, if nematodirus is known to be a problem, dosing the kids in May will remove the risk of disease. The danger periods are illustrated in fig. 5.3.

Jan.	Feb.	Mar.	Apr.	May.	Jun.	Jly.	Aug.	Sep.	Oct.	Nov.	Dec.
			■								

5.3 The danger period for nematodirus infestation

BASIC PRINCIPLES FOR THE CONTROL OF GUT ROUNDWORMS

1. Where roundworm parasites are concerned, sheep and goats are affected by the same ones and thus they should be considered together.
2. All species of worm behave in the same way and thus can be considered as one type for practical purposes (expect for nematodirus).
3. Control is aimed at providing the goats with a clean pasture and if possible removing them from dangerous pasture.
4. Drugs given to goats to rid them of parasites (anthelmintics) should ideally be used to prevent the goats from putting out eggs on to the pasture, thereby keeping down the challenge. Throughout this chapter 'dose' means to treat the goat with an anthelmintic (*see* page 89).
5. If possible, graze goats (and sheep) on pasture one year and graze another species (cattle or horses) on it the following year because the worms that affect cattle and horses are different from those affecting sheep and goats. This will ensure virtually clean grazing for each species.

SUGGESTED WORM CONTROL STRATEGIES

I. The Commercial Herd on a Large Acreage
(a) The 'ideal' situation where pastures can be alternated each year with cattle or horses or an arable crop
● Dose goats after kidding (i.e. the day following kidding).
● Continue to graze them as long as the weather allows. Be prepared to dose in September if necessary. Dose at housing.

(b) The commercial herd where no resting of pasture is possible
Pasture grazed by goats in the previous autumn. Hay or silage aftermaths available.
● Graze goats until the end of May or June. Dose and move in June when an aftermath becomes available. Dose at housing.

II. The Small Herd Grazed Continuously on One Paddock
● Dose after kidding

- Dose all animals (excluding kids below 6-8 weeks) every 4-6 weeks from mid May to the end of September.

III. The Verge-grazed Goat

- Ensure that the goat is moved continuously and avoid grazing the same area the following year. After a year's 'rest' the ground will be free from parasites, so grazing should be in a 2-year cycle. No dosing is necessary in this situation.

IV. The Stall-fed Goat

The goat is probably free from parasites but equally she is all the more susceptible, because she has poor immunity. Parasite control will depend upon what is fed and where the forage comes from. The following points may be of value:

- Browse can be considered to be free from parasites.
- Hay can be considered to be virtually free from parasites.
- Grass cut and fed from grassland where goat-house manure is spread *may* be dangerous. The danger can be reduced by composting the manure before spreading.

Symptoms of worms

Symptoms of worm parasites may be either very mild or very severe. The type of symptom may also vary with the age of the goat, adults tending to be more resistant than kids. Symptoms may also depend upon the type of worm involved: stomach worms such as Haemonchus are blood suckers and they tend to cause anaemia. Intestinal worms such as Trichostrongylus cause irritation and hence symptoms of diarrhoea. Normally many species of worm are present at the same time and so a mixture of symptoms are noticed. Goats suffering from a light challenge may show a loss of body weight or, in young kids, simply no increase in weight. More obvious changes may be noticed with heavier burdens including 'staring' coats, reduced appetite and perhaps diarrhoea. Lactating does may have milk yields below their expected performance. Goats with very severe worm problems are generally in poor bodily condition, weak, anaemic and have diarrhoea. In cases of acute (i.e. sudden) heavy challenge, animals previously in good condition may show rapid decline and become weak over a matter of days. Diarrhoea would probably be a marked feature.

Treatment

Tremendous strides have been made in recent years in the field of Anthelmintics (*see* below). Modern preparations are safe and extremely effective. Even goats suffering from very severe disease can be cured by dosing with these drugs. Care should be taken with very sick goats; it may be prudent to divide up the dose and administer it over two days.

ANTHELMINTICS

These drugs are 'Anti-helminths', active against helminth worms. There are many compounds available and your veterinary surgeon will give you guidance as to which one to use. The modern drugs are extremely effective at getting rid of the worms from the goat's stomach and intestines. Some of them have the added bonus of controlling lungworms and liver fluke. Because they are so very effective there is in fact very little to choose between them and it is more fruitful to concern oneself with when they should be given and working out grazing control strategy.

Giving Anthelmintics
Administrations can be either by injection, drenching, in feed, or paste. The injectable forms carry some risk and are, therefore, best avoided. Goats are extremely 'fussy' eaters and drenching tends to

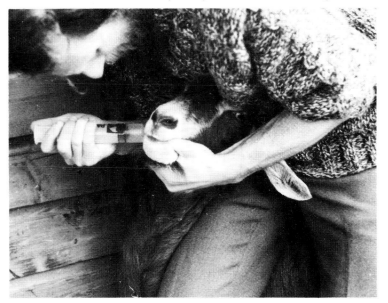

PLATE 5.1
Dosing the goat

be the most successful – but not the easiest – method of administration. Hold the goat's muzzle with fingers in the mouth between the front and back teeth. Slowly pour the medicine, tilting the nose upwards. A plastic bottle or an old teapot can be used. Dosing goats is not an easy task and you should be prepared for spillage. Your veterinary surgeon will supply you with a large plastic syringe for dosing purposes. Some of them, intended for the administration of anthelmintics, have a long spout which is ideal for drenching purposes.

Some authors recommend the paste forms of medicine which can be applied to the tongue. Care must be taken with dosage because these are often formulated for horses. Read the manufacturer's instructions carefully because some drugs have a milk-withholding time when milk should not be used for human consumption.

Myths and Misunderstandings about Worm Control

Paddock Rotation
Some years ago scientists attempted to combine together efficient grazing management and worm control. The concept of a series of six paddocks was proposed which the goats grazed in rotation, spending about one week in each paddock. Although sensible from a grassland management point of view, it can be disastrous as a method of parasite control.

Stocking Rate
Once a certain minimum number of animals is reached for an area of grassland, parasites become a potential danger. Thought must be given to parasite control and it makes little difference whether two or twenty goats are grazed there because the number of infective larvae that can result from just one goat grazing a paddock is very high. It is the timing of exposure to the parasite that is important. If the goats are all moved from the pasture at the right time, then no serious challenge or disease results.

Resistance to Anthelmintics
Although this problem has been recorded in sheep, as yet there is no evidence of it arising in goats. Simply changing one's anthelmintics would avoid the problem if it arose.

Anthelmintics used For Goats
*Indicates the drug is active against parasite

| Drug | Trade names | Activity spectrum | | | | Remarks |
		Gut roundworms	Tapeworms	Lungworms	Fluke (adult only)	
Benzimidazoles — Thiabendazole	'Thibenzole'	*				—
Fenbendazole	'Panacur'	*	*	*		—
Oxfendazole	'Systamex'	*	*	*		—
Albendazole	'Valbazen'	*	*	*	*	—
Haloxon	'Loxon'	*				Occasional neurotoxicity reported
Levamisole	'Nilverm'	*		*		Toxicity reported

Although all these drugs are widely used in goats, only Fenbendazole is licensed for use on goats in the U.K.

Worm-free Goats

Outside the laboratory, it is neither practical nor desirable to have 'worm-free' goats. A healthy goat is one that experiences a small amount of challenge but avoids the danger periods when the pasture is heavily contaminated with larvae.

The Effects of Worming Goats

The administration of anthelmintics to goats removes the population of worms established in the goat at that time. If the goat is then returned to infected pasture she immediately starts to take in infective larvae. Thus another population of adult worms is established within her, in the space of about 4-6 weeks.

Worm Egg Count

In the past, parasitologists have attempted to quantify the weight of parasite infestation by counting the parasite eggs present in a gram of faeces. Results expressed in the form of say '500 eggs per gram' were thought to indicate more serious infestation than one of '100 eggs per gram'. Unfortunately this assumes a constant rate of egg production by the worms in the goat and there is no basis for this assumption. This does not mean to say that dung samples are not reliable in confirming that parasites are present in animals.

Infestation in Indoor Reared Kids

Kids are extremely susceptible to worms, and on account of their habit of soiling feed containers it is possible to spread larvae via contaminated feed. Always ensure that feed and water containers are kid-proof and cannot be contaminated.

LUNGWORM

The symptoms of lungworm disease are coughing goats which, in severe cases, fail to thrive. The worms inhabit the air passages and cause inflammation (parasitic pneumonia). Two species of worm can be responsible for these symptoms, *Dictyocaulus filaria* and *Muellerius capillaris*. Both worms are present in the United Kingdom and the United States. *M. capillaris* is much more widespread than *D. filaria*.

Dictyocaulus filaria

This worm can infest both sheep and goats, but there is generally little disease associated with it in goats. This worm has a similar life cycle to the gut roundworms, except that the adults inhabit the lungs, not the intestines. Thus there is an extra stage of migration by the larvae from the intestines to the lungs, via the bloodstream. The life cycle of this worm is illustrated in fig. 5.4.

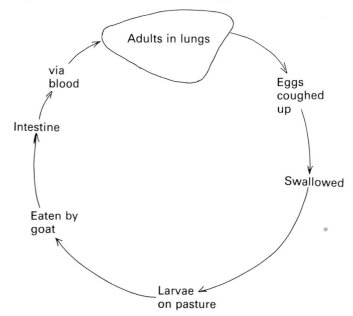

5.4 The life cycle of the lungworm, D. filaria

Muellerius capillaris
Although this is a parasite of sheep and goats, it rarely causes disease in sheep but can cause disease in goats. In Britain many goats are thought to be infested by this parasite, but this does not necessarily mean that they are suffering disease. Disease symptoms include coughing and increased breathing rate. The life cycle of this worm is complex and includes an intermediate host such as a snail or a slug. These are generally found on wet pastures.

The Patterns of Lungworm Disease
Disease associated with both worms can be considered together. Coughing is generally seen in late summer or autumn. Because the parasites are picked up while the goats are grazing, disease is generally seen at pasture. It may also be encountered in stall-fed goats when forage is fed from fields previously spread with goat or sheep manure.

Control

The principles of control are similar to those for gut roundworm. Challenge arises mainly from the pasture in early summer, resulting from both overwintered larvae and larvae from infested adults. (In the case of *M. capillaris* it is the snail or slug.) Therefore goats moved and dosed as suggested for gut worm control should avoid lungworm disease. Development of immunity against lungworm infestation is much better than it is against gut roundworms. This probably accounts for the relatively low prevalence of disease. Most goat owners can afford to be relatively unconcerned about lungworm disease in goats for the following reasons:

1. Infestation with *D. filaria* is relatively rare, and it generally causes little disease.
2. Grazing control measures applied for gut roundworms should also control lungworms.
3. Some anthelmintics used to control gut roundworms are also effective against lungworms; they include Albendazole and Oxfendazole.

LIVER FLUKE DISEASE (FASCIOLIASIS)

The fluke is a parasite of the livers of ruminants such as the goat and sheep. Adult flukes inhabit the bile ducts of the liver and they can remain there for months or years. An intermediate host, the mud snail (*Limnaea truncatula*), is necessary for the completion of the life cycle of the fluke. The snail only lives in wet, poorly drained areas and so liver fluke disease tends to be prevalent in the high-rainfall, western regions of the United Kingdom.

The Life Cycle

A diagrammatic illustration of the life cycle of the liver fluke is shown in fig. 5.5.

The adult fluke is a hermaphrodite possessing both male and female organs. It lays eggs which pass into the bile and out with the goat's droppings. Hatching of the eggs occurs if the temperature is above 10°C and miracidia are released. The miracidia swim about in search of a snail and if they are successful they penetrate the tissues of the snail.

Several stages of development take place within the snail and after a minimum period of about five weeks cercariae are released on to the pasture. The cercariae swim in the moisture on the pasture finally settling on blades of grass. They then become covered by a secretion which hardens and they remain on the grass as metacercariae. The metacercariae are ingested, along with the herbage, as the goat grazes.

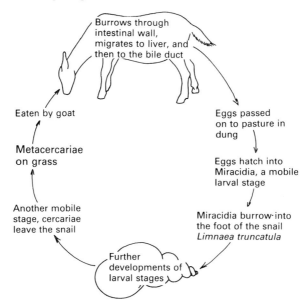

5.5 The life cycle of the liver fluke, Fasciola hepatica

The Pattern of Fluke Disease

As with the other parasites, the weather conditions have a profound influence upon the disease and the amount of disease varies from year to year. The amount of challenge to the goat is dependent upon the survival of the mud snail. Thus the factors that influence the survival of the snail determine the disease threat. The two key factors are temperature and rainfall. A temperature of 10°C is essential for the developement of the snail, and adequate rainfall is necessary. These two criteria are generally present in the summer in the wetter parts of Britain but to varying extents. It is possible, however, to measure these two parameters over the early summer months and then produce a forecast of the severity of the challenge that can be expected that autumn and winter. This is the 'fluke forecast' carried out each year by the Ministry of Agriculture, Fisheries and Food. Goatkeepers can use this advice in order to avoid disease in their goats.

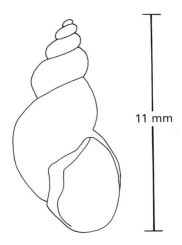

11 mm

5.6 Shell of Limnaea truncatula

The Use of the Fluke Forecast

If the year has been wet, then the time that the goats are liable to take in parasites is from late August through to December. Thus, all pastures can be safely grazed from spring until the middle of August, irrespective of how many snails are present. If the fluke forecast is 'bad' the goats should be moved to well-drained pasture which is unlikely to have any fluke present. On the other hand a 'good' fluke forecast means that there is going to be little challenge that year and such precautions need not be taken.

Jan.	Feb.	Mar.	Apr.	May.	Jun.	Jly.	Aug.	Sep.	Oct.	Nov.	Dec.

5.7 The danger periods on pasture where fluke is a problem (in a 'bad year')

Control

In theory fluke can be controlled by three main methods:
1. Eliminating the snail.
2. Preventing the goat ingesting the parasite.
3. Eliminating the fluke in the goat's liver and thereby reducing the contamination of pasture.

1. Eliminating the snail

This is achieved by *either*:
- drainage, *or*:
- the use of molluscicides such as copper sulphate.

Both methods are expensive and have to be carried out thoroughly, the use of molluscicides also has an environmental contamination risk. For the majority of goatkeepers with only a few animals they are not worthwhile from an economic point of view.

2. Preventing the goat ingesting the parasite
This method means obtaining the MAFF fluke forecast and planning grazing accordingly. It is difficult for goatkeepers who have no alternative grazing to use. The stall feeding of goats on hay, greengrocers' trimmings or other forage may be an answer in bad years. This can only be done if the economic loss from the disease in terms of lost milk or dead animals is greater than the extra feed cost incurred.

3. Eliminating the fluke in the goat's liver
The use of drugs active against fluke in the live animal (flukicides) not only removes the parasite from the liver but also reduces the number of fluke eggs that are deposited on to the pasture. This in turn reduces the challenge to goats in subsequent grazing years. It would seem sensible for goatkeepers who graze fluke-infested pastures to use Albendazole as their standard anthelmintic. This wormer has the added bonus of removing adult flukes from the liver.

Symptoms
Liver Fluke disease can manifest itself in either of two forms, acute or chronic.

Acute fluke disease
Eating massive numbers of the parasite over a short time results in large numbers of immature flukes passing from the gut of the goat into the liver. Acute disease results from the damage caused by the tremendous sudden injury to the liver. The symptoms are sudden and death can occur in a very short time. In less severe cases the goats are dull, off their food and may kick at their abdomens in pain. If post-mortem examination is carried out, the liver is seen to be enlarged, bloody and friable.

Chronic fluke disease
More common than the acute form, chronic disease is seen in the late winter. Affected animals become progressively weak and lose weight. Diagnosis is confirmed by the demonstration of fluke eggs in the droppings of the goat.

Treatment
Various drugs are available to treat this disease, though none of them are licensed for use in goats. Rafoxanide (Flukanide) is a drench form and Nitroxynil (Trodax) is an injectable preparation. Albendazole (Valbazen) is an anthelmintic which is also effective against fluke but only in the adult form.

TAPEWORMS

Many species of tapeworm can inhabit the small intestine of goats. They include *Moniezia expansa* which is found throughout the world. This particular species has a mite as its intermediate host, the goat becoming infected by eating mites while grazing.

The Significance of Tapeworm Infestation
To my knowledge little research work has been carried out in order to investigate the harmful effects of tapeworm infestation. On the surface, one would assume that harm may be caused but there is little evidence to support this hypothesis in sheep, where the problem has been studied. My opinion is that tapeworm infestation in adult goats is of no significance but may occasionally be important in kids.

Diagnosis and treatment
Diagnosis of the disease is made by examination of the goat's droppings. In the unlikely event of problems the anthelmintic Albendazole (Valbazen) can be used for the treatment of affected kids.

OTHER DISEASES AND EXTERNAL PARASITES

Grass Tetany (Hypomagnesaemia/Grass Staggers)
See Chapter 6, 'Problems Associated with Milking Animals', page 115.

FOOTROT

This is a specific lameness of grazing goats and sheep, commonly seen in the summer months and occasionally during the winter, in housed animals. Typically infection enters the hoof at the skin horn junction (*see fig. 3.3*) and causes inflammation of the sensitive laminae. The infection spreads below the hoof causing severe pain and lameness. Because it affects both sheep and goats it is relevant to point out the widespread distribution of this disease in sheep flocks.

The bacterium involved and predisposing factors
All cases of footrot are associated with infection by *Fusiformis nodosus*. The predisposing factors to the infection are damp, lush pasture, typical of those found in the United Kingdom. The moisture of the pasture has a tremendous bearing on the distribution of the disease. It is noticeably absent from the arid regions of the world. The effect of the moisture is to soften the hoof horn, allowing the bacterium to penetrate more easily. *F. nodosus* can persist in the hooves of infected animals for years, but in the pasture it can only live for about a week. This is very important if you are trying to avoid the problem, and I shall discuss this under 'Control'.

Symptoms
Lameness is slight at first and there is only a small amount of swelling between the claws. As the horn is underrun and the sensitive laminae become involved lameness becomes increasingly severe. Typically there is a foul smell associated with it and the horn begins to separate from the underlying tissue. The animals are reluctant to walk and may graze on their knees. Milk yield may fall and some animals will lose weight.

Treatment
Affected goats should receive attention in the form of hoof paring, in order to remove the underrun horn. The next step is to apply antiseptic agents in order to remove any infection. These are best applied topically on to the affected area, but in severe cases your veterinary surgeon may prescribe antibacterial drugs, by injection.

Requirements for the Treatment of Footrot

- *hoof shears;*
- *a sharp knife;*
- *local antiseptic, either 5% formalin or 10% copper sulphate or a suitable spray;*
- *a hard surface such as concrete.*

After treating the feet the goats should be left to stand on a hard, dry surface for as long as possible, in order to allow the antiseptic to work. You should not turn them out into a muddy field. When the job is completed clean the equipment in 5 per cent formalin (or its equivalent) and burn the hoof trimmings. If any of the herd are affected then the whole herd should be put through a footbath (*see* page 54). As goatkeepers are well aware, goats dislike water and forcing them through a footbath is not an easy job. Foot bathing should not be carried out more than once a week because it is detrimental to horn and skin.

Control

1. Attempting eradication by pasture management
Because *F. nodosus* only survives for a maximum of ten days on pasture it is possible to eliminate the infection from fields. This can be achieved by carrying out the treatment as detailed above, on say three occasions, with an interval of ten days.

The goats must not be returned to the same pasture until ten days have elapsed.

2. Vaccination
There is a commercial vaccine available for sheep, which has been used in goats called Clovax (Wellcome Foundation Limited). No work has been carried out to test its effectiveness in the control of footrot in goats, but I feel that it is worth while using in problem areas. The primary course of immunisation comprises two injections separated by an interval of eight weeks. The manufacturers recommend this should be started in October. Single booster injections should be given twice-yearly in October and February each subsequent year. In-kid goats should be vaccinated six weeks before kidding or two weeks afterwards.

PINK EYE (CONTAGIOUS OPHTHALMIA, NEW FOREST EYE)

As the name suggests, this problem is an eye infection that spreads in herds of goats (and sheep). The conjunctiva or membrane that covers the eye becomes inflamed. In severe cases the cornea, lying below the conjunctiva, is involved.

Season of the Year
This is a disease of the summer when the agents that spread the germ abound. It can be spread by flies, dust and long grass. At colder times of the year, feed troughs are probably responsible because they cause crowding of the goats.

Symptoms
At first a 'watery' eye is noticed with excess tears spilling over on to the skin. There may be some reddening and the cornea becomes cloudy. Over the course of a few days the discharge thickens and

becomes sticky. Recovery generally occurs within a fortnight, but very occasionally, ulceration of the eyeball may occur, with loss of its fluid.

Treatment
Some cases will recover without any treatment but your veterinary surgeon may supply you with an antibiotic ointment to aid healing and reduce the risk of spread to other goats. Application of the ointment should be made several times a day.

Control
The organism generally associated with this condition is *Rikettsia conjunctivae*. *Neisseria ovis* has been recorded in the problem abroad. Recent work in the United States has revealed that a mycoplasm, called *Mycoplasma conjunctivae*, may also be involved. These agents are present in the eyes of some goats. They act as reservoirs of infection for other goats when conditions are right for spread. Thus the only control measures that can be applied are to treat affected animals quickly, isolate them and avoid 'crowding' goats when the disease is present in the herd.

FLUORIDE POISONING (FLUOROSIS)

Grazing pastures contaminated by fluoride can cause poisoning in goats. In the United Kingdom, the most common situation is that of pastures that surround brickworks. The fumes from the brickmaking process deposit fluoride on to the pasture. Certain parts of Bedfordshire are renowned for this contamination. The symptoms include lameness and stained teeth. There is no treatment, and in such situations goatkeepers would be advised to run a 'flying herd', that is, to buy in milkers each year.

EXTERNAL PARASITES: FLY STRIKE (MIASIS) MAGGOT INFESTATION

Any open wound in the summer months attracts flies. Certain flies, the blowflies, actively seek such wounds in which to lay their eggs.
    After hatching, the larvae of the blowfly eat the tissues of the unfortunate goat, causing intense irritation and discomfort.

Treatment
Larvae should be removed and the wound cleaned. A dressing incorporating an insecticidal preparation such as Negasunt should then be applied, The wound should be examined twice a day until healing has occurred.

Control
Examine and dress all skin wounds, being careful to inspect for signs of fly eggs, or larvae. Remove all predisposing causes of wounds such as projecting nails. Treat diarrhoea promptly in the summer months because this attracts the flies.

The use of sheep dip
The impregnation of the coat of the goat with insecticides may be worth considering, especially for non-milking goats at pasture. With sheep flocks, dipping is the rule. For a goatkeeper with small numbers of animals, the showering of sheep dip, using a watering can, would be satisfactory. If sprays or dips are used for lactating goats, care must be taken to read the instructions. Some of them require that milk be discarded as it is not fit for human consumption. Organophosphorus compounds such as Asuntol (Bayer) can be used if the goats are treated immediately after one milking. The milk should be fit for use at the next milking.

WARBLE FLIES (HYPODERMATOSIS)

Warble flies can affect goats overseas; I have seen no reports of cases occuring in the United Kingdom although they may rarely occur. *H. lineatum* was found in goats in Turkey, and this species is present in Britain. The condition is characterised by the appearance of swellings under the skin of the back.

THE NASAL FLY (*Oestrus ovis*)

The larvae of this fly spend part of their development in the nasal cavities and sinuses of sheep and goats. The sheep is the specific host but occasionally goats may be affected. In the United Kingdom the fly is only active in the summer months and it is not generally a widespread problem. In hotter parts of the world, the fly is active for longer periods and one South African survey revealed that three-quarters of goat herds were infested.

Life cycle
The adult fly deposits its larvae around the nose of the goat and the larvae migrate into the nasal cavities. After several weeks they pass into the sinuses, and from there on to the ground. After pupating on the ground the adults mate and commence to deposit larvae.

Symptoms
The adult flies upset the goats while they are grazing. The larvae

irritate the linings of the nose causing the goats to sneeze and produce a discharge.

Treatment
I have no experience of treating this condition but various organo-phosphorus compounds are reported to be effective. The treatment is generally given by mouth.

Ticks and Tick-borne Disease

Ticks are large parasites that feed periodically on sheep and goats, dropping off from the animal when they have taken a meal of blood. Although unsightly, their main damage is to spread disease when they bite the goat. In problem areas such as the south-west of the United Kingdom, routine dipping or spraying of goats may be advisable. Alternatively, the odd tick can be treated individually by applying insecticide locally. It is best not to pull off the tick because the mouth parts often remain in the skin.

Louping-ill (Encephalitis)

Louping-ill is a viral disease that causes an inflammation of the brain of sheep. It is transmitted by the bite of a tick infected with the virus. Very occasionally it may cause disease in goats, though there are no reports of clinical cases. A survey carried out on feral goats in Scotland revealed that many of the goats had antibody to the disease. This suggests that the disease probably occurs from time to time.

The symptoms in sheep are those of fever, convulsions and paralysis. It is only likely to occur in hill areas where ticks abound.

Orf (Contagious Pustular Dermatitis)

This is a highly infectious viral disease of goats and sheep. It causes scabby sores around the lips and muzzle. The disease can pass from sheep to goats and vice versa. Although rapid spreading, it is generally not a serious disease except that it does result in loss of productivity. Whereas in sheep flocks it predominantly affects young lambs, in goats it seems to cause problems in all ages of stock.

Symptoms
Initially, swellings and pustules occur which soon become scabby and crusty. They are especially common at the corners of the mouth. Most cases are fairly mild, the scabs become dry and fall off

so that the wound is healed over in about three weeks. Loss of condition occurs because the goats are disinclined to eat on account of the pain. The yield of milking does may fall. Grazing animals tend to suffer more than goats fed quantities of concentrate food.

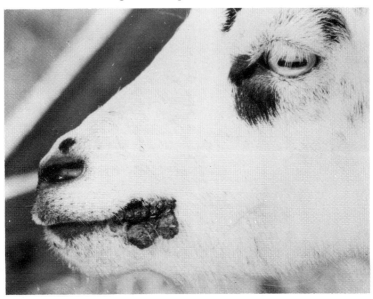

PLATE 5.2
Symptoms of orf on a goat's mouth

Transmission of orf
Spread occurs by direct contact with other animals or through inanimate objects such as fences and feed troughs.

Treatment
It is doubtful whether treatment helps at all but the application of creams and ointments may reduce the spread of orf around the herd. Offering some concentrate food stuff will help to avoid any setback in grazing goats.

Control
Isolation of affected goats may reduce the spread of the disease. Vaccination may be of value in the early stages of an outbreak. The vaccine generally comes in 50-dose packs and is only economic for goatkeepers with several animals. The vaccine is given by scratching

the skin with a special applicator, either inside the thigh or preferably under the tail.

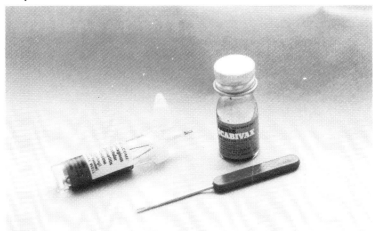

PLATE 5.3
Orf vaccines

PLATE 5.4
The administration of orf vaccine

The virus can persist for many years in the scab and therefore outbreaks may occur after several years of freedom from disease. It is probably best not to use the vaccine unless problems have occurred with this disease in former years.

Human infection

Humans can become infected and painful sores develop in the skin. Care should be taken when handling affected animals or when administering the vaccine.

GID, STAGGERS (COENURUS CEREBRALIS INFESTATION)

If the larval stage of the dog tapeworm, *Taenia (Multiceps) multiceps,* develops in the brain, this disease may result. As the cyst grows it causes pressure and damage to the brain cells of the goat.

Symptoms

The symptoms seen depend on the part of the brain involved. Affected goats may circle, show a high-stepping gait, or blindness in one eye. If the cyst lies in the spinal cord, posterior paralysis may result. Occasionally, a softening of the skull may be noticed where the cyst lies beneath the bone. It may be necessary to use X-rays to confirm that the cysts are present in the skull.

The cycle of infection

Infected dogs put out tapeworm eggs in their faeces. Goats grazing contaminated herbage eat the eggs and the larvae migrate to the brain. Dogs become infected by eating uncooked carcasses which contain tapeworm cysts.

Control

1. Cook sheep and goat offal before feeding it to dogs in order to destroy the tapeworm cysts.
2. Administer tapeworm tablets to dogs that come in contact with goats and sheep.

Treatment

Treatment of the goat will rarely be possible and slaughter will probably be necessary.

JOHNE'S DISEASE (PARATUBERCULOSIS)

Johne's disease is a chronic incurable infection of the intestines of goats, sheep and cattle. It is caused by *Mycobacterium johnei*, a slow-growing bacterium belonging to the same family as tuberculosis. Infection causes a thickening of the intestine with the consequence that less nutrients are absorbed.

Symptoms
Loss of flesh is the main symptom, the goat gradually becoming emaciated. The droppings may become slightly loose but diarrhoea is rare.

Treatment
There is no treatment for Johne's disease; affected animals should be slaughtered. Diagnosis is made by laboratory examination of the goat's droppings.

Control
The important factors about this disease are:
1. The organisims can survive on pasture for up to a year.
2. Infected animals may put out bacteria on to the pasture for months before any symptoms are noticed.
3. There are no reliable tests to pick out infected animals before they show disease symptoms.

Thus control on an infected farm is extremely difficult. The best method would be to slaughter all stock and not keep goats, sheep or cattle for between one and two years. If that is not possible, vaccination may be used but this is regulated by the Ministry of Agriculture, Fisheries and Food. Your vet will be able to obtain the vaccine from them. Studies in Greek herds suggest that vaccination is very effective in controlling the disease in infected herds. In order to avoid early transmission, kids would be best reared away from their dams. Separation at birth is advisable and feeding them on milk replacer instead of dam's milk.

Avoidance of Johne's disease
Because Johne's disease is so difficult to eradicate from a premises, all goatkeepers should try to avoid purchasing infected stock. It is always worth enquiring from the vendor of a goat whether or not he or she has had any animals with the condition. It is also important to remember that infected cattle can infect goats.

PHOTOSENSITIVITY

In certain circumstances goats exposed to several hours of sunlight may develop changes in the skin, which can vary from slight thickening to very severe damage. Photosensitisation occurs only when certain photodynamic substances are present in the skin. In the presence of sunlight an allergic type of reaction occurs, resulting in a thickened skin and dermatitis. This condition only occurs if the goats are white or have white patches of skin, the explanation for this being that they are unpigmented. Photosensitisation should not be confused with sunburn which usually appears some time after exposure to sun.

Substances predisposing to photosensitisation
Many plants that goats eat have a propensity to cause this disease directly, for example St John's Wort (*Hypericum perforatum*). Other plants can cause problems indirectly, if the liver is damaged or not working efficiently. The breakdown products from these plants (phylloerythrins) accumulate in the tissues of the body instead of being excreted. Such plants include the lupin (*Lupinus angustifolius*), and other plants giving similar problems include kale (*Brassica rapa*) and lucerne (*Medicago sativa*).

Symptoms
Thickening occurs on the white (unpigmented) parts of the coat especially around the ears and muzzle. White goats, such as the Saanen are particularly prone to this condition. The skin thickening is especially noticeable on the backs of the goats and to a lesser extent on the sides. In very severe cases the skin is shed. The affected goats may show signs of irritability and excitement.

Treatment
Once removed from the direct rays of the sun the goats soon recover. Putting them in a shed for a week or two should be all that is required in mild cases. If the skin is very badly damaged then your vet may prescribe drugs to reduce the inflammation and control any infection that may develop. By removing the goats from the pasture and putting them inside, one normally prevents the goats having access to the plants involved.

PINE (COBALT DEFICIENCY)

Pine is a wasting disease caused by deficiency of cobalt in the feed. Cobalt is essential in very small quantities and hence its designation as a trace element. The condition is found mainly in areas deficient in cobalt. Pine is relatively uncommon in British goats; I have never seen it, but other authors have described it. The low prevalence of pine is accounted for by the facts that first, only certain soils are low in cobalt and second, the majority of British goats receive adequate attention and mineral supplementation. An important point in the pattern of the disease is the fact that cobalt cannot be stored by the goat and must be continuously available in small quantities.

Symptoms
Cobalt is essential for the production of vitamin B12 in the rumen. The vitamin B12 requirements of ruminants such as goats are quite high. Deficiency causes an inability to metabolise propionic acid and a failure of appetite. This loss of appetite results in loss of weight and eventually death.

Treatment
Affected animals respond well to the administration of either cobalt or vitamin B12. Cobalt sulphate is generally given, the usual amount being 10 mg by mouth, at weekly intervals. For convenience cobalt 'bullets' can be administered to adult goats. These lodge in the reticulum and slowly release cobalt to the goat. Because the reticulum is underdeveloped in young animals the bullets are not satisfactory for kids.

BLOAT (RUMINAL TYMPANY)

All goats above weaning age possess a large fermentation chamber called the rumen (*see* Chapter 4, page 63). Carbon dioxide and methane gases produced from the fermentation are eliminated by the animal belching. If for some reason the gases cannot be got rid of, pressure builds up in the rumen. The left-hand flank of the goat becomes distended due to the enlarged rumen and the animal has great difficulty breathing. It is fairly uncommon in goats compared with cattle and sheep.

Bloat can result from many causes, including:

1. An obstruction of the oesophagus (*see* CHOKE).
2. Paralysis (as in TETANUS).
3. Eating foods which produce lots of gas, over a short period of time.
4. 'Frothy' bloat, when the gas bubbles will not break down easily.

By far the most common type is 'frothy' bloat seen when a stable type of gas bubble is formed in the rumen. When this occurs the gas remains as a 'froth' in the rumen defying the goat's attempts to belch it up.

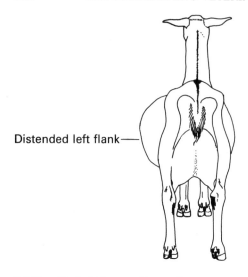

Distended left flank

5.8 The effect of bloat on the goat's appearance

The reasons why the gas bubbles tend to foam or froth is very uncertain. Many theories have been put forward and many explanations may be partially correct. The high surface tension of the bubbles can be partially explained by the composition of the food material eaten. For example, lush pasture with a high protein content has been cited. The quantity of saliva produced may also be important, because the saliva aids the breakdown of the froth. Occasionally, bloat develops in goats suddenly fed large quantities of concentrates.

Recognising ruminal bloat
Affected goats have distended left flanks. They may show signs of discomfort; kicking, bawling or grinding their teeth. In more serious forms, the goats have obvious difficulties in breathing, because the abdominal contents restrict the space into which the lungs expand.

Treatment
Preventing the animal eating anything is the first step to take. The type of treatment then depends upon the seriousness of the condition. In very mild forms, relief can be given by drenching (carefully!) with 50 ml of vegetable oil such as peanut oil. Ten ml of washing up liquid could be used in an emergency. Proprietary silicone-based drenches are also available. Alternatively your vet may pass a

stomach tube in an attempt to release the gas. In acute cases more drastic action is required and the rumen may have to be punctured with a knife or preferably a trocar and cannula. This is inserted by a stabbing action with the trocar inside the tube of the cannula. Once in place the trocar is withdrawn to allow the gas out (*see also* page 180).

PLATE 5.5
The trocar and cannula

After-care
Because treatment can interfere with the delicate rumen functioning, 'seeding' the stomach with fresh faeces or live yoghurt may be beneficial. Feeding branches from fruit trees such as apple may help to improve rumen movement.

Prevention of bloat
Take care when grazing goats on rapidly growing pasture. Restrict the time of grazing to short periods, especially if lucerne is grazed. The goats may be drenched with vegetable oil before they are turned out on to pasture. Never turn hungry animals out on to rich pasture; fill them with hay before they graze for the first time in the spring.

HEAT STRESS

Heat stress can result when goats are exposed to strong sunshine in the summer months. It is especially likely to occur in tethered goats that are unable to seek shade. The symptoms are panting, rapid breathing and thirst.

Treatment
Move the goat into the shade as quickly as possible and offer them water to drink. If necessary, dowse the backs of the animals with cold water.

Chapter 6

PROBLEMS ASSOCIATED WITH MILKING GOATS

Two groups of problems are commonly encountered in goats producing milk:

1. metabolic disorders;
2. diseases of the udder and teats.

They conveniently form a natural division for this chapter.

METABOLIC DISORDERS

Metabolic disorders result from a disturbed input-output balance of nutrients. At the start of lactation, milk production represents a sudden drain on the goat's reserves. Constituents such as protein, fluids and salts, are all lost from the body and the goat's metabolism has to cope with those sudden changes. Years of selection for high-yielding goats has resulted in animals capable of giving far more milk than is required by twins or even triplets. Thus the strain on the goat is made worse and it is hardly surprising that sometimes things go wrong.

Milk Fever (Hypocalcaemia)

This problem occurs in milking does at or around kidding, but it is fairly uncommon. A survey in 1964 revealed only five cases out of nearly a thousand animals. Milk fever is most likely to be seen in the age range of 4-6 years. In a study of forty goats with hypocalcaemia in Norway the time of onset was:

17% less than one week before kidding
25% during kidding and next few days
20% less than three weeks after kidding
37.5% more than three weeks after kidding.

The problem is essentially a disturbance of calcium metabolism. Calcium is abundant in the bones and teeth, with smaller quantities present in the blood and other tissues. The mineral is vital for muscle contraction, and if blood levels fall this function of enabling muscles to move, ceases.

The Biochemistry of Calcium

A basic understanding of the turnover of this mineral will help the reader to understand why milk fever occurs. Fig. 6.1 is a diagrammatic representation of the movement of calcium within the lactating doe.

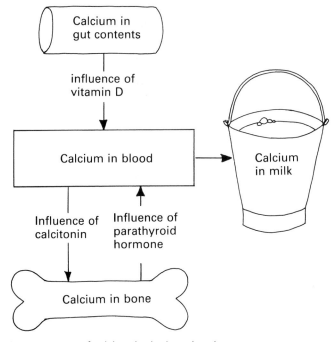

6.1 The movement of calcium in the lactating doe

In the non-lactating doe some of the calcium present in the diet is absorbed into the blood and transferred to the bones for storage. Vitamin D influences this movement, together with the hormone calcitonin. When required, calcium can be released from the bones back into the blood under the influence of parathyroid hormone. Problems arise in does around kidding, when the situation suddenly changes and calcium has to be brought back out of the bones into the blood.

Why Milk Fever Occurs

Hypocalcaemia simply means 'low blood calcium'. This does not mean that the goat is deficient in calcium, the bones are full of it. It simply means that temporarily the blood calcium level has fallen because the complex control mechanism described in fig. 6.1 has failed. At kidding the goat is suddenly called upon to produce in excess of 2-3 litres of milk each day. Milk is high in calcium and this obviously represents a heavy loss, which temporarily lowers the blood calcium level to less than half its normal value (from 2.48 to 0.94 millimole/litre).

Symptoms

Because calcium is essential for muscle tone, goats affected with milk fever appear to be weak. They generally lie down and become unable to rise. Cudding ceases and they do not usually pass water or faeces. If left untreated, the goat would probably become gradually weaker and eventually slip into a coma and die.

Milder forms may be quite common, being responsible for protracted kiddings. Symptoms of slight weakness (ataxia) affecting the gait, may also be noticed in about half of the cases. The temperature of the affected goat is frequently below normal (38°C).

Treatment

Response to the administration of calcium salts is good, the symptoms normally disappearing within a few hours. Calcium borogluconate is the commonly used injection which can be given either under the skin or directly into a vein. In acute cases, the best treatment is probably to give 25 ml of the 40 per cent solution into the vein and also inject a further 75 ml under the skin. This technique combines a rapid response with a deposit of calcium which can be slowly absorbed from under the skin.

After the injection of calcium borogluconate, the goat's body normally adjusts and mobilises its own calcium from the bones. If the doe is lying on her side, then she should be propped up with a bale so that she is lying on her breastbone (brisket). This prevents rumen fluid entering the lungs. For the following two milkings it is best not to take off too much milk (as this will impose more strain on her).

Prevention of milk fever

Understanding the biochemistry of calcium enables us to realise what goes wrong and to rectify any faults. If calcium is fed in high quantities just prior to kidding, the control mechanisms register this high input. They send adequate quantities into the bone but also reduce the absorption of calcium salts from the intestine. If this mechanism persists after kidding, the calcium lost in the milk depletes the reserves, resulting in a state of low blood calcium. Conversely, if a diet high in phosphorus and low in calcium is fed for the week prior to kidding, the goat's metabolism is ready to adjust to the sudden change. In this situation, calcium is already being mobilised from the bones and milk fever does not occur. It is not easy, of course, to formulate the ideal low-calcium diet unless one consults a book on nutrition chemistry. As a rule of thumb only investigate this further if cases of milk fever are occurring in your herd. Farmers whose cows have a known susceptibility to milk fever use vitamin D just before calving or even dose the cow with calcium gel at calving. There are no reports of their use in goats but they could be used in problem herds.

KETOSIS (ACETONAEMIA)

This condition can occur within the first month of lactation. It is very similar to pregnancy toxaemia (*see* Chapter 4) and only occurs in housed, lactating animals, in the winter. Ketosis is basically the inability of the milking doe to keep up her potential for milk production, due to faulty management. Goats have been bred for very high milk yields, this inherited attribute being termed 'genetic potential'. The doe can only yield that quantity of milk if her nutrition and management are just right. If the doe yields to her potential on an unsatisfactory diet, then something has to give and in this case ketosis develops. Ketosis results from any one of a variety of management errors, or a combination of them (*see* ACETONAEMIA, page 71).

HYPOMAGNESAEMIA (GRASS TETANY OR GRASS STAGGERS)

The word hypomagnesaemia means the condition of low blood magnesium. In normal animals a certain concentration of magnesium salts is present in the blood. When these are not present, symptoms of 'staggers' may be seen. It is mainly a problem of adult milking goats; it is not common.

Symptoms
The onset of symptoms can be very rapid. The goat behaves strangely, becomes unco-ordinated and 'staggers'. On closer examination the muscles can be seen to be contracted and the animal is excitable.

The predisposing factors
Ruminants are unable to store magnesium for long periods and the disease is an input-output problem. This is best explained by referring to fig. 6.2.

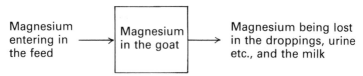

6.2 Input and output of magnesium in the milking doe

The goat attempts to keep the magnesium in the blood constant but this is affected by variations either in intake of the food or in the output, for instance in the milk. Many factors can influence intake of magnesium in the diet, the obvious one being a change of diet. Three situations generally predispose to hypomagnesaemia, they are as follows:

1. Lactating does grazing rapidly growing spring grass;
2. Heavy lactation;
3. Stress from poor weather (for example, no shelter in the winter).

Treatment
The administration of magnesium salts can completely reverse the course of this disease. They must be given in time, however, as the disease can be rapidly fatal. Magnesium sulphate is normally given under the skin, but occasionally your vet may administer some into the vein. Response to treatment can be fairly rapid, the goat being back to normal the following day. The injection of magnesium sulphate can precipitate a lowered blood calcium. It is therefore prudent to inject calcium borogluconate at the same time as the magnesium sulphate.

Prevention
As the disease occurs mainly in lactating animals, prevention is generally easy because they are handled twice a day. Ensuring that magnesium is present in their concentrate rations is the most simple method. If extra is required, calcined magnesite can be fed at 6 grams a day. Other methods available are the provision of licks or even giving the goats a 'bullet' which remains in the stomach. The bullet is impregnated with magnesium and if given by mouth, settles in the reticulum, where it gradually releases magnesium.

DISEASES OF THE UDDER AND TEATS

Any impairment of function of the udder of the dairy goat is very serious. It is also not surprising that such diseases are fairly common, if one considers the following three points:

1. Dairy goats have been selected for milk yield over the centuries and a goat with a large exposed udder has been produced.
2. They are subjected to twice-daily milking which imposes a strain on the defence system of the teats.
3. Goats are frequently milked one after another so that the possibility of transmission of germs from one udder to the next is very high.

Mastitis

Mastitis is an inflammation of the udder, almost always associated with germs and showing a great range of variation from mild to very severe. In the dairy cow, where the disease has been studied extensively, it has been demonstrated that inapparent or sub-clinical mastitis is very common. This can also be shown to occur in the goat. Thus, mastitis may be affecting the glands of goats when the owner does not even realise it! Mastitis is economically important because it reduces milk yield and it may also cause permanent damage to the udder.

Recognition of Mastitis — the Symptoms

1. *Sub-clinical*   No visible changes in the milk or udder. Simple tests can demonstrate changes in the milk. Experiments have revealed a lowering of yield by affected halves.
2. *Mild*   Clots in the milk.
3. *Acute*   Swollen, painful udder. Clots and watery milk, reduced yield. The goat may be off her food. There may be blood in the milk. It is generally seen close to kidding.
4. *Very Acute*   The changes as in 3; almost invariably blood is present, the udder may even feel cold and clammy. The goat is not eating and is a very sick animal, possibly too weak to rise. Her temperature is high. Gangrene of the udder may be the final outcome. It is generally seen close to kidding.
5. *Chronic*   Repeated episodes of 2.

It should be noted that these observations are for guidance only and, of course, the goat may start with a mild mastitis which may progress to the acute form.

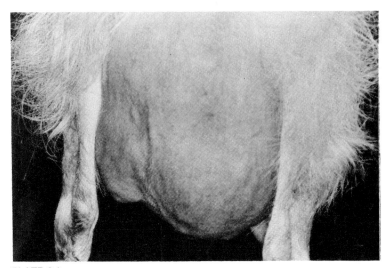

PLATE 6.1
Clinical mastitis

How Common is Mastitis?

There are few figures available for the prevalence of mastitis in goats. One survey revealed that in large herds twenty-five per cent of goats can have one infected half. One Indian report recorded 32 cases out of a flock of 140 goats. In an investigation in Greece, nearly half the udders were found to be infected. A chronically infected herd in Norway contained 28 goats with mastitis out of 50. Many of these cases were, of course, inapparent or sub-clinical mastitis. In a small survey carried out by the author, only five cases of clinical mastitis were recorded in 231 adult goats over a year (not all of them were lactating). A 1964 United Kingdom survey revealed that clinical mastitis was responsible for large losses.

In summary, I conclude that sub-clinical mastitis is common and perhaps more common than is realised.

Fig. 6.3 illustrates the basic anatomy of the udder. The two halves are separated by a septum of fibrous tissue, which means that infection from one side does not spread to the other side.

The Spread of Mastitis

Micro-organisms, especially bacteria, are always present in an attack of mastitis. They enter by way of the streak canal, ascend the teat duct and invade the glandular tissue. Because these bacteria are incapable of moving on their own accord something must allow them to gain entry to the teat. Thus, there are two important aspects of mastitis: the presence of bacteria and, more importantly, the factors which allow them to invade the teat. The two main sources of infection are:

1. The environment of the goat (including the bedding and the contaminated external surface of the udder and teats);
2. Other goats with mastitis.

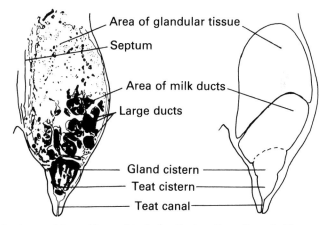

6.3 *Anatomy of the udder and teats (vertical section of one half)*

Factors Predisposing to Mastitis
The factors which allow the bacteria to invade are thought to be many and varied and the situation is very complex. Below is a list of such factors, although many of them are disputed or not proven.

1. *Physical injury*
   - Bushes
   - Fences
   - Butting by other goats
2. *The Milking Process*
3. *Overmilking*
4. *Machine Milking*
   - Incorrect vacuum pressure
   - Teat cup liners
   - Vacuum fluctuations
5. *Housing and bedding*
6. *Teat sores.*

In summary, it is known that the above factors do influence the cases of mastitis but scientists are frequently at a loss to explain how they have this effect.

The Effects of Mastitis on the Gland

Once the bacteria have invaded the gland they begin to multiply very rapidly, because they are in an ideal warm moist medium. The udder attempts to defend itself by producing defence cells which eat up (phagocytose) the bacteria. Frequently this defence system is enough to overcome the infection. Many such cases of sub-clinical mastitis are eliminated, without the goatkeeper ever knowing that they occurred. If, on the other hand, the weight of infection is too great, the bacteria continue to multiply and their presence inflames the cells of the gland, destroying them or changing the nature of the milk they produce. In very acute cases the inflammation is so great that blood and pus appear in the milk. Goats with very acute mastitis develop a temperature as a result of the inflammatory process in the udder.

Diagnosis of Sub-clinical Mastitis

Several indirect tests are available to diagnose this problem. They are cheap, simple to use and results are available almost immediately. They are based upon chemicals which demonstrate the presence of white blood cells, the existence of which indicate that mastitis is affecting the udder. One study revealed that the Modified Whiteside Test was the most accurate for use in goats. To carry out this test, one drop of sodium hydroxide is added to five drops of the milk to be tested, on a glass tile. If, after twenty seconds, a viscid mixture with clumps is seen, sub-clinical mastitis is present. Your veterinary surgeon will advise you on these tests.

Cell counts

Cell counts are sometimes used in mastitis investigations. They are a quantification of the cells demonstrated in the tests described above. Essentially, a high cell count (over 500,000 cells per ml) is indicative of high levels of sub-clinical mastitis, whereas a low cell count suggests that mastitis is not a problem. Again your vet will help you to interpret the findings. Further research on the interpretation of the cell counts is urgently required. Some authors suggest that only anything greater than 1 million cells per ml is significant. It is fair to conclude that our knowledge of the value of cell counts in goats is very unsatisfactory. Recent work by the MMB has borne this out.

Infective Agents Associated with Mastitis

Many different types of bacteria can be isolated from cases of mastitis in goats. The following list stresses the fact that many organisms can be associated with caprine mastitis.

Streptococcus agalactiae, S. dysgalactiae, S. pyogenes, S. zooepidemicus, Staphylococcus aureus, Staph. epidermidis, Staph. pyogenes, Yersinia pseudotuberulosis, Enterobacter cloacae, Pseudomonas aeruginosa, Clostridium perfringens type C, *Corynebacterium ovis, Bacillus cereus, Klebsiella pneumoniae, Mycoplasma putrifaciens.*

In most outbreaks it is rather academic to know which type is involved for two reasons:

1. Treatment should be initiated early on, before laboratory test results are available.

2. We are aware that the types of bacteria in the goat's environment are legion and the circumstances which allow them to gain entry are more important than knowing which organism happened to invade the gland.

Treatment

Mild, and even more acute cases, respond well to therapy with antibiotics, infused directly into the affected half. It must be realised, however, that permanent damage in the form of destroyed gland tissue may have occurred. In such cases, the gland may well secrete less milk in subsequent lactations. Very acute cases may not respond well and the infection may even kill the goat. Sometimes the episode will be so severe as to make the doe useless as a producer of milk, even though she recovers.

Milking does

Infection during lactation is helped by the fact that the milk flushes out the organism but is hindered by the fact that the antibiotic used as treatment is also flushed out.

Dry does

The reverse applies for cases occurring in dry does (*see* page 130).

Preparations

All the preparations for the treatment of mastitis (called intramammary tubes) are designed for use in the dairy cow but they can be used in goats. There are a plethora of them available containing a wide range of different drugs.

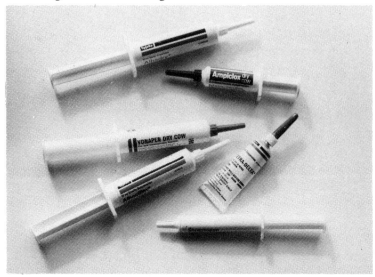

PLATE 6.2
Intramammary tubes

In the absence of specific recommendations for goats it is probably best to use a cow dose, that is one tube for each half that is treated. On no account give in to the temptation to economise by using half a tube for each half, as you may spread infection from one half to the other via the intramammary tube. Your veterinary surgeon will prescribe for you enough repeat infusions to establish a cure. The number of repeats is determined by the type of drug used and the base which it is in, being either quick or slow release. There is an obvious dilemma in treatment between obtaining a satisfactory cure and attempting to avoid the waste of milk. Milk must be discarded because it contains drugs. Observe the instructions given on the tube as to how long the milk must be discarded after the last treatment. Administration should be undertaken as hygienically as possible. The milk should be stripped out before doing so. Clean the end of the teat with cotton wool soaked in alcohol (spirit) or antiseptic. Insert the nozzle into the teat, squeeze out the contents and then try to massage the ointment up the teat, sealing the end with the finger and thumb. In goats with tiny teat apertures it may be necessary for your vet to use a cat catheter in order to put the antibiotic into the teat.

After milking, clean teat with cotton wool and spirit

Insert the nozzle fully into the teat canal, apply steady pressure on plunger to deliver the full dose

Hold the end of the teat and massage the teat, moving the antibiotic upwards into the quarter

Insert the teat in teat dip

6.4 Administration of an intramammary tube

Taking Samples for the Laboratory

Sometimes it is beneficial to take milk samples for examination at a laboratory. This is done in the case of a serious herd outbreak or in order to check the suitability of an antibiotic drug that has been prescribed. Cleanliness is essential if the sample is to be of any value, and the following steps should be observed. A badly taken sample may be contaminated and cause more confusion than no sample at all.

1. Obtain a sterile container.
2. If teat is dry, remove dirt by wiping.
3. Discard the first stream of milk into another container.
4. Clean the end of the teat with cotton wool soaked in spirit (e.g. methylated spirit); wait for the spirit to dry.
5. Discard the next few streams of milk.
6. Collect the sample.

7. Label and date the sample, then send it to your veterinary surgeon.

PREVENTION OF MASTITIS

For as long as people milk goats, mastitis will always occur. Just treating the symptoms, that is clinical cases, will only solve part of the problem. It is necessary to avoid sub-clinical mastitis, which is very prevalent, but not noticed. This means preventing new infections.

As vaccines are of no use in this role, new infections must be prevented either by **better husbandry** or by **breeding**. Unfortunately, there is little prospect of dairy goats being selected for mastitis resistance. If this type of breeding is pursued, other, less favourable characteristics may be produced or good traits lost. It is also probably the case that the high-yielding, easy-milking goat, is the most prone to mastitis. She is also, of course, the goat most breeders are striving to produce. Perhaps in the long term there is a case for selecting goats with moderate yield, capable of producing from forage, instead of from the heavy concentrate diets that are used at present.

A. The Control of Mastitis by Better Husbandry Designed to Prevent New Infections

In the dairy cow, mastitis control by this approach has been very successful. The same principles apply to goats and I know goatkeepers who have achieved success by using better hygiene and husbandry.

Preventing New Infections

This can be achieved in many ways. On page 119 the predisposing factors to mastitis were listed. Each point will now be covered.

Physical injury to the teats by bushes, fences etc.
The dairy goat does not have a lot of ground clearance and thus attention must be paid to detail to avoid mechanical damage to the udder and teats. Door steps and bedding retaining boards can cause injury.

Butting of udders
Play and aggressive behaviour may result in udder damage. Sometimes this can be avoided by keeping groups of goats separate.

The milking process

This should be carried out quietly, efficiently and hygienically. The doe becomes conditioned to being milked and therefore helps in the process by 'letting down' her milk. In fact milk let-down is not a conscious act but is a reflex that follows stimulation of the teats and even the familiar stimuli of, for example, seeing the milker. The pathway of the reflex is partly transmitted by the nerves and partly by the action of the hormone oxytocin. The oxytocin acts by squeezing the alveoli which contain much of the milk and sending it down into the gland cistern. Myoepithelial cells actually contract and squeeze the alveolus.

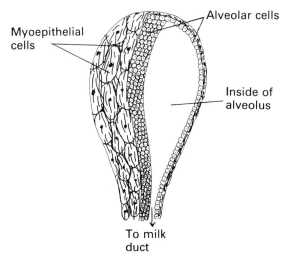

6.5 *The alveolus and myoepithelial cells*

Overmilking

It seems obvious that excessive milking of goats will result in damage to the teats. This has been a widely held theory in dairy cows but not conclusively proven by experiment. Avoiding excessive stripping would seem a sensible precaution, however. If the goat is milked to excess, then very slight damage to the teat canal results, which predisposes to bacteria entering. Research has shown that infection mainly takes place immediately after the milking; thus any steps taken to reduce stress to the teats at this stage is beneficial.

The importance of a regular routine at milking

The importance of all this to mastitis is that this milk let-down reflex is prevented by fear, excitement, stress or pain. For example, change of routine at milking or the presence of a noisy visitor tend to inhibit the reflex and thus prevent efficient emptying of the udder. This means that the doe has to be milked for a long period in order to obtain the milk and this causes damage to the teats.

Hygiene

The importance of being clean at milking cannot be over-emphasised for two reasons. Firstly, to avoid contamination of the teats and secondly to assist clean milk production. Remember that most infections occur immediately after milking and thus the cleaner the teats are, the smaller the chance of infection. Udder cloths should be disposable and not used for one goat after another. In the opinion of many, including the author, visibly 'clean' teats in goats, say, at pasture, need not be washed before milking.

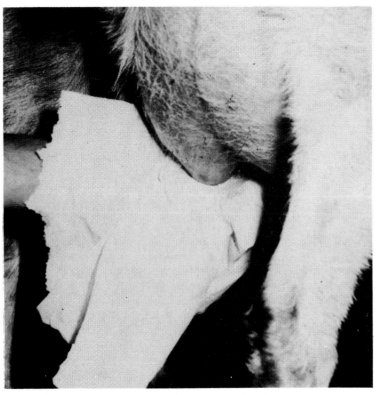

PLATE 6.3
Udder washing

The use of teat dips

Because most infections take place immediately after milking, the application of antiseptic to the teats just after milking has been shown to prevent new infections. The teats are immersed in teat dip as soon as they have been milked and this temporarily sterilises them. This gives the teat time to recover from the milking process and to regain its protective function.

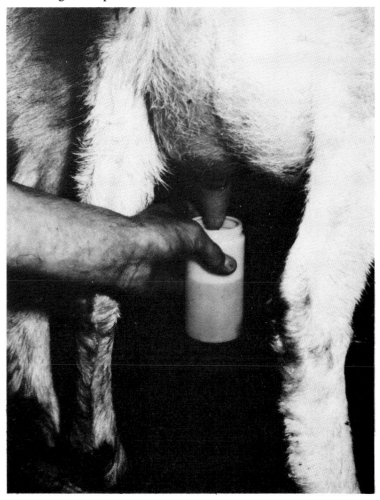

PLATE 6.4
Teat dipping

Machine milking
There are several aspects of this milking method which predispose to mastitis. Firstly, damage can occur to the teats as a result of using a machine. This can arise from faulty vacuum pressure, excessive vacuum fluctuation or worn teat cup liners. The machine should be checked regularly to ensure that these faults can be corrected.

The second aspect about the machines is that they can physically carry infection from one doe to the next. Clinical cases of mastitis should always be milked last and the machine carefully sterilised afterwards.

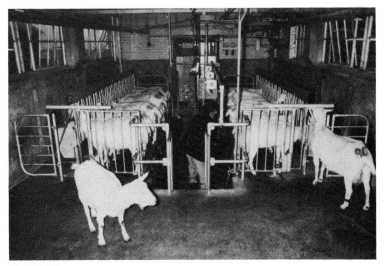

PLATE 6.5
The milking parlour at the National Institute for Research into Dairying

Housing and bedding
Bacteria capable of causing mastitis are around the goat all the time. It is obvious that the cleaner we keep them, the fewer the bacteria that will be present on the teats. Of equal importance is the state of the bedding as regards physical damage to the udder and teats. Cold, wet concrete chills udders; clean straw bedding does no harm at all.

Teat sores
Cuts, chaps and sores on the teats not only cause problems at milking because the goat is in pain but they also provide hiding

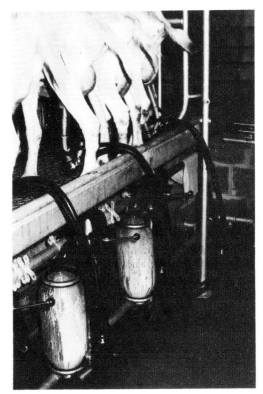

PLATE 6.6
Machine milking at N.I.R.D.

places for bacteria. Care should be taken to avoid such sores and prompt treatment should be given in the form of udder creams. More serious, infected sores require the application of ointments containing an antibiotic. In herd problems, your veterinary surgeon should be consulted, in order to identify the cause and to suggest ways of control.

B. Prompt Treatment of Existing Infections
By treating existing infections quickly, the goatkeeper reduces the reservoir of infection that poses a threat for the other goats in the herd.

Dry Doe Therapy
This practice is widespread in the dairy cow. At drying off all udders

are infused with antibiotics which persist in the udder for several weeks. The rationale for its use in cows is that many infections carry on from one lactation to the next and cause severe disease just after calving. The dry period is a good time to eliminate the infection because no milk has to be thrown away and the antibiotic remains in the udder for a long period. This practice can be applied to goats in just the same manner. I have strong reservations about using antibiotics in this way because it is no logical answer for the long term. In my view greater emphasis should be placed upon good husbandry and breeding for goats with sensible yields. These goats are less prone to mastitis. However, it may be prudent to use dry doe therapy in goats that have had mastitis during the previous lactation.

The Sale of Goat's Milk
There are few regulations about the sale of goat's milk compared with the mass of legislation about the sale of cow's milk. Goatkeepers do have the moral obligation to sell wholesome, fresh milk that is fit for human consumption. Never sell milk from goats with mastitis because raw goat's milk is often fed directly to infants. Some bacteria associated with mastitis are a human health hazard if drunk. Milk containing antibiotic residues (from does being treated for mastitis) must never be sold. Some people are allergic to certain antibiotics and drinking contaminated milk can make them very ill. Antibiotics in milk may induce resistance in bacteria and thus reduce the effectiveness of antibiotics when given to humans for medical purposes. Always take careful note of the milk-witholding time given on the mastitis treatment.

Cuts and Lacerations of Teats

These can be very important and very frustrating problems to deal with. Very serious cases should be referred to your veterinary surgeon immediately. The most serious ones, frequently torn on barbed wire, are those that penetrate the teat canal. This can be easily seen if milk is leaking from the wound. Mastitis may develop if infection gets into the gland through the wound. From a practical point of view, milking a goat with a cut on its teat is difficult and time-consuming, for the obvious reason that the wound is painful. Unfortunately if the goat is to continue to produce milk then she must be milked. This action, of course, delays the healing of the wound. Very severe cuts will be stitched up by your vet but superficial ones may respond to the application of sticking plasters.

Your veterinary surgeon will almost always prescribe a course of antibiotics, if the wound includes the teat canal. This is to safeguard against mastitis developing.

Prevention

Avoid putting goats into areas that contain objects capable of damaging their udders. Clear up bits of barbed wire and check fences regularly to ensure they are goat-proof. Try to avoid situations where goats will be tempted to jump through barbed wire fences — for example hungry goats or goats in season.

Oedema (Dropsy) of the Udder

Some does develop very swollen udders just before kidding. If the udder is of normal temperature and contains no abnormal hardness then oedema is most probably present. To test for this, press the thumb into the skin and if a depression or pit remains, then oedema (or fluid) is present in the udder. This is a natural phenomenon and is normally of no significance. If the doe is in great pain when it comes to milking her, then massaging may help. Should the problem persist your veterinary surgeon may prescribe an injection to remove some of this fluid. Normally the fluid will pass away in time but one study described a method of incising the skin to release the fluid. I would never resort to such treatment and I certainly would not suggest that goatkeepers try it.

Blood in the Milk

This occurs from time to time, the milker noticing pink milk. It usually follows a blow to the udder. In newly kidded does this finding is more frequently a symptom of a ruptured blood capillary. At kidding the udder frequently becomes congested and with a build-up of pressure of milk, small blood vessels occasionally burst. The blood leaks into the milk, turning it pink. Normally the vessels cease bleeding within a short time, heal up and no more is seen. The milk is aesthetically unpleasant and must not be sold. Assuming that no mastitis is present, it could safely be fed to dogs and cats. If the haemorrhage is very severe, call your veterinary surgeon and he will give injections to arrest the haemorrhage. He will probably insert a long-acting antibiotic into the affected half and instruct you not to milk it for a few days.

Fall in Milk Yield

This is a common symptom of many diseases, especially ones that make the doe go off her food. It cannot be considered alone and as it is an important symptom, you must seek professional help from your vet.

Goat Pox

Two types of goat pox occur, a mild form and a very severe (malignant) form. The malignant form is not present in the United Kingdom but affects goats in Africa and Asia. Benign goatpox causes raised papules and pustules on the skin of the udder and teats. It can also affect the lips and mouth. The pox may persist for five or six weeks, gradually becoming 'crusty' as time passes.

Treatment
Applying ointment to the sores may help to reduce spread and also encourage rapid healing. There are many suitable udder creams available which contain an antiseptic such as cetrimide or chlorhexidine. The ointment should be removed by wiping all the teats with a disposable towel, before milking. Affected goats should be milked last and the equipment sterilised in order to attempt to reduce the spread of pox.

Male Milkers

Occasionally bucks may commence to produce milk and mastitis may result.

Supernumerary Teats

Some kids are born with more than two teats. These teats are best removed at a young age so that no scars remain when the doe comes into milk. They can be snipped off using curved scissors and the wound dressed with antiseptic powder or cream. The British Goat Society discourages goatkeepers from breeding from goats affected by this condition.

Blind Udders

Occasionally does will kid and produce an udder one side of which will not milk, even though the milk-secreting tissue is present. The problem stems from a milk duct that is not open. Reports indicate that the condition is inherited. Sometimes careful surgery will enable the teat to be opened up to allow milking of the half. Any such interference, however, carries with it a risk of mastitis developing.

WARTS ON TEATS

Warts are benign tumours of the skin. Frequently they are associated with a virus. Generally they disappear after several months as immunity develops.

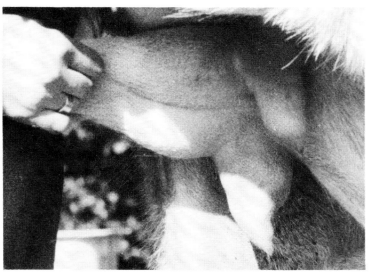

PLATE 6.7
Buck with glands producing milk

Chapter 7

PROBLEMS ASSOCIATED WITH BREEDING, PREGNANCY AND BIRTH

BREEDING

Season

In northern latitudes goats are seasonal breeders, only capable of mating at certain times of the year. In the tropics does come on heat all the year round. In the United Kingdom the season extends from September to February.

The Reproductive Cycle

Several hormones are involved in the delicate functioning of the oestrous cycle. They are listed below with their abbreviations and their sites of production.

Hormone	Abbreviation	Site of production
Follicle stimulating hormone	FSH	Anterior pituitary
Oestrogen	—	Follicle in ovary
Luteinising hormone	LH	Anterior pituitary
Progesterone	—	Corpus luteum in ovary
Prostaglandin	PG	Uterus

At the start of the breeding season the cycle proceeds as illustrated in fig. 7.1.

The stimulus of decreasing day length causes the release of follicle stimulating hormone from the anterior pituitary. This causes the development of follicles in the ovary. The follicles then release oestrogens which result in the doe exhibiting the symptoms of oestrus or heat. The circulating oestrogen acts upon the hypothalamus/pituitary to reduce FSH and to stimulate the production of LH. This results in ovulation and an egg is released from the now mature follicle in the ovary. A summary of this process is given in Fig. 7.2.

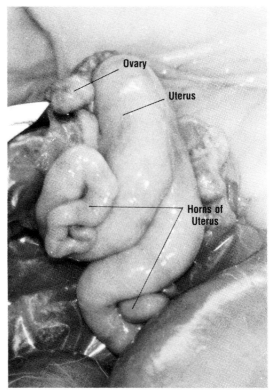

PLATE 7.1
The uterus and ovaries of a non-pregnant doe; the knife point
is on one ovary

The corpus luteum forms from what remains of the follicle after
ovulation. This starts to produce progesterone which is necessary to
maintain pregnancy. All this takes place in about one week, com-
mencing a few days before the heat is noticed. If pregnancy does not
interrupt the cycle, then prostaglandin is released from the uterus.
This hormone is 'luteolytic' and causes the corpus luteum to disappear. Once this is gone so too is the progesterone and the whole
cycle starts once again. Should pregnancy result during the cycle
then prostaglandin is not released and the corpus luteum remains.

Signs of Heat (Oestrus)
Does on heat become vocal, calling loudly. The vulva swells slightly
and a clear mucus discharge may be noticed. A great deal of tail

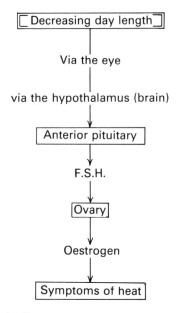

7.1 *Chain of events leading to oestrus*

wagging occurs and the area around the tail looks characteristically 'dirty' or 'wet'. Very rarely the does may show mounting behaviour, the doe on heat being mounted by another doe.

Length and frequency of heat
This is extremely variable. I have seen heats as short as twelve hours

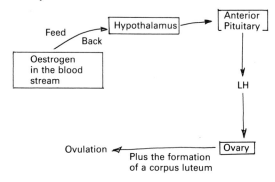

7.2 *The hormonal action leading to ovulation*

and as long as forty-eight. Other authors give between 36-96 hours and 13-27 hours, and one colleague reports a four-hour heat in his own doe. Within the season goats come on heat approximately every 19-21 days.

PLATE 7.2
Doe on heat standing to be mounted by another doe

MATING

A doe in heat will normally stand to be mounted, although a nervous doe in strange surroundings may fidget and move. The mating is quick. The male mounts, penetrates and after a few thrusts throws back his head when ejaculation occurs.

ARTIFICIAL INSEMINATION

Trials are being carried out in the United Kingdom using artificial insemination in goats. The technique using 'raw' semen is relatively easy but the semen has a very limited lifespan of a few hours. Frozen

PLATE 7.3
Normal mating

semen is being used in the present trials. This may promise an easy
alternative for goatkeepers who live a great distance from the buck
of their choice. Only does that have kidded previously are suitable
for use with AI.

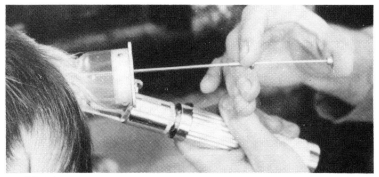

PLATE 7.4
Artificial insemination, the speculum allows the operator
to locate the cervix

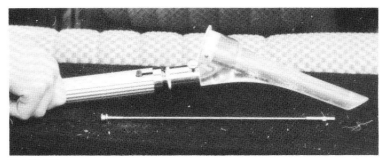

PLATE 7.5
Insemination equipment

EXTENDING THE BREEDING SEASON

The constraint of the natural breeding season can be partially overcome by artificially bringing does into heat. This is best attempted in the months just before the season normally commences. Thus, July or August may be suitable in the United Kingdom. The advantages of doing this are to have does kidding early on in the year. This extends the time during which milk is produced in quantity.

Procedure
1. Progesterone-impregnated sponges are carefully inserted into the vagina of the doe using an antiseptic cream to prevent the introduction of infection (Veramix sponges containing 60 mg of medroxyprogesterone or Chronogest [Intervet] are suitable) It may be easier to insert them using a clean, well-lubricated finger, rather than the applicator. This is especially true in maiden females. The sponge should be positioned about halfway along the vagina, not too close to the cervix.
2. 500 i.u. of PMSG (pregnant mare serum Gonadotrophin) are injected 19 days later.
3. The sponges are withdrawn 17-21 days after insertion (they have a little 'tail' of nylon cord by which they can be pulled out). Other goats will sometimes nibble the cord and pull the sponge out by accident.
4. Heat should occur 12-36 hours after the sponge is withdrawn.

Results
The success rate for this technique is generally in the order of over 50 per cent pregnancies. The nearer to the natural breeding season that this technique is carried out, the greater is the success rate. It is

difficult to determine which does have conceived because the non-pregnant does resume seasonal anoestrus. They do not come on heat until the normal breeding season has commenced. If no pregnancy results the goatkeeper has lost little and the doe is served, as normal, when she is on heat. One practical drawback of this technique is that the sponges come in packs of fifty. Your veterinary surgeon may be prepared to split up packs and insert sponges into individual does in the smaller herd. The technique should not be attempted less than 150 days after kidding, because the lactation will reduce chances of success.

The Influence of Light
The major factor in the initiation of the breeding season in goats is decreasing day length. The natural stimulus of the shorter days of autumn can be mimicked by using light-proof buildings. This technique works experimentally and there is no reason why it could not be used by goatkeepers. Does are put into a building where light can be controlled. They are then given decreasing 'days' by shutting doors and windows progressively earlier.

Identifying Breeding and Mating Problems

The broad types of problems that may present themselves are listed below. Having identified the type of abnormality, the reader can refer to the relevant section in order to try and determine the cause.

1. Does that fail to come on heat.
2. Unmated does that come on heat and then cease to cycle.
3. Does with heats of irregular length.
4. Does with regular heats after mating (repeat breeders).
5. Physical problems of mating.

1. Does that Fail to Come on Heat
Three important questions should be posed before complex investigations are initiated. The most important question to ask oneself is, of course, Is there a problem? Could the heats be occurring without being noticed? Heats may persist for very short periods of even just a few hours. Is it possible that she is pregnant? This is especially relevant if the doe has been brought in from elsewhere.

Is the goat an intersex? This is only likely in a maiden goat. Such animals are always infertile (*see* INFERTILITY DUE TO INTERSEX, page 144).

Non-lactating does that fail to come on heat
Failure of does to come on heat, between, say October and November, must be treated as a problem. Often the trouble is that the doe is either too thin or too fat.

Goats in poor bodily condition may be undernourished. In these instances increase the total food ration, especially of energy food such as cereals. Observe whether cycles commence and if they do not, ask your veterinary surgeon for advice. He will probably give the doe an injection of prostaglandin to bring her into heat. One and a half ml of Lutalyse (7.5 mg) PGF$_2$ will normally stimulate oestrus within a few days; 1 ml of Estrumate (ICI) is reported to be equally effective. If some underlying disease is responsible for the goat's condition, your veterinary surgeon will advise you on this.

On the other hand, excessive fatness may affect a goat's cycles. Again seek your vet's advice; an injection of prostaglandin will probably bring her into heat. After she has been mated try to get her into lean condition, otherwise she will be too fat at kidding.

Milking does that fail to come on heat
Does kidded for less than three months: does come on heat while they are still milking but the lactation may suppress the oestrus (in the first part of lactation). If the doe has kidded within the previous three months, leave her alone and wait to see if heats recur naturally.

Does advanced in lactation: does that are milking well may fail to come into heat if their diet is poor. Increase the energy food intake of these does for three weeks. If no heat occurs, consult your vet who will probably bring her on by an injection of prostaglandin.

2. Unmated Does that Come on Heat and then Cease to Cycle
In my experience, these goats can often be made to start cycling again by using prostaglandin injections, providing that it is not too late in the season (i.e. before February).

The cause may be either movement to new surroundings or possibly ovarian cysts.

3. Does with Cycles of Irregular Length
Goats show a wide variation in length of the oestrous cycle. Different authors quote varying lengths; for example three sources quote the following:

17-23 days
18-21 days
19-21 days.

It is probably safe to assume that anything outside the 17-23 days is 'abnormal'.

Short cycles or continuous heat
At the beginning of the breeding season some goats come on heat twice within a week; they then settle down to normal-length cycles. Some does exhibit frequent heat periods or they appear to be in a continuous state of heat. This is normally an ovarian problem where cysts develop in the ovary. It is as if the ovary becomes 'stuck' in the heat phase of the cycle; oestrogens are produced and they are responsible for the signs of heat.

The condition can be treated, unless the symptoms have progressed to the stage of masculinisation, where the doe begins to behave as if she were a buck. In these cases the animal is best slaughtered. (Some goatkeepers with large herds retain these masculinised animals because they can be usefully employed to indicate other does that are on heat.)

Treatment
Hormones can be given by your vet to change the state of the ovary. He would probably use chorionic gonadotrophin (Chorulon Injection, Intervet Laboratories Ltd) which contains two hormones: LH (luteinising hormone) and FSH (follicle stimulating hormone). These hormones will encourage the old cysts to disappear and will also stimulate new follicles to develop.

This condition is fairly common; for example in one study of dead goats at abattoirs, 24 cases were seen in a total of 1,020 goats. It must be remembered, however, that some of these goats may well have been slaughtered because of this condition; thus it may have been a biased sample.

Long cycles
Occasionally, long cycles may be detected in does. If they are multiples of 20 days, e.g. 40 days, then it is possible that they are in fact normal cycles and the owner has missed observing a heat

period. In other cases early embryonic death may have occurred; that is, the doe conceived to service, but the embryo died. In this case the doe might go for some time before she has her next heat.

4. Does with Regular Heats After Mating (Repeat Breeders)
There are a number of reasons why a doe may return to service at regular intervals after being served:

● The buck may be infertile. This can normally be checked by finding out if other does served by him have conceived. It is also possible to check a sample of his semen in order to clarify the situation.
● Does can show heat during a pregnancy. The signs of oestrus are usually less well defined. If in doubt have the doe served, it should normally do no harm.
● If the doe is milking then this may be detrimental to her conceiving (high yielders). This is very rare because normally the peak of lactation (three weeks after kidding) is in the spring or early summer. As the breeding season is in the autumn, unless induced artificially, the doe should be well on in lactation. Theoretically it is a stress on the doe to support an embryo and to carry on yielding a great deal of milk. If this is suspected, increasing the energy ration at service may help to get the doe to conceive.
● The service is being carried out at the wrong time. Possibly too early on in the heat period.

5. Physical Problems of Mating
Occasionally, everything is theoretically right in that the doe is on heat, the buck is with her but successful mating does not occur. The majority of problems in this situation would lie with the male.

The buck
Reluctance to serve may be due to painful hind legs and feet. Arthritis seems to be a fairly common problem of stud bucks. It is probably caused by excessive feeding of calcium and keeping them on cold, damp, bedding. Lameness from other causes should also be considered. (*see* Chapters 3 and 8.)
    Overwork: only when severely underfed and undernourished, will a buck refuse to mount, but attention should be given to the buck's nutrition. During the breeding season bucks often go off their food and it may be useful to vary the diet in order to try to tempt them to eat.

INFERTILITY DUE TO INTERSEX

The progeny of polled (naturally hornless) goats are frequently infertile. It is therefore best not to rear these animals for breeding purposes. As a prevention policy always mate goats which are polled or of unknown status to a horned goat.

PREGNANCY AND KIDDING

The average length of pregnancy is 150 days but there is obviously some variation about this mean. It is not abnormal for goats to kid four days either side of 150 days. Breed differences account for some of the variations.

Tests for Pregnancy
The goatkeeper would find it useful to know whether or not the doe has conceived. Several different methods are available and these are described below.

Absence of heats
In many species such as the cow, the fact that the served cow does not come on heat again is a fairly good indication of pregnancy. Alas, some goats do exhibit heat symptoms during pregnancy and therefore this indication, although useful, is not fool-proof. Of course, does with a 'cloudburst' (pseudo-pregnancy) do not come on heat.

The milk test
The Milk Marketing Board will carry out a test on the milk of a lactating goat in order to test for pregnancy. Formerly, the progesterone test was used at twenty-four days but this gives rise to an unacceptable number of false positives. The present test can only be performed at fifty days and it involves the demonstration of oestrone sulphate. Goatkeepers should contact the MMB for sample containers; the cost is £1.00 per sample (January 1982). Occasionally, after having a positive test, something may happen to the foetus late on in pregnancy so that no kid is born. The drawback of this test is that the doe must be milking and it cannot, therefore, be used on goatlings in-kid for the first time.

Ultrasonic examination of the abdomen
A probe is appled to the abdominal wall of the doe and the examiner listens for evidence of the sound pulse being bounced back off a kid,

if it is present. It can be used from about the sixtieth day of pregnancy.

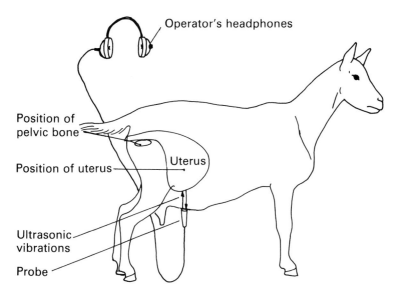

7.3 Position of probe for ultra-sonic pregnancy examination

I have used this in a few sheep and goats with only limited success. It is probably fair to conclude that it is good in the hands of a very experienced user. Other workers conclude that it gives a proportion of false negatives; that is, does are declared to be empty when in fact they are pregnant.

The rectal pole technique
Originally designed for sheep, this test can be applied to goats and Guss recommends its use in does between 70 and 110 days of pregnancy. In essence the pole is inserted into the rectum in order to allow the examiner to feel the pregnant uterus more easily. The examination would seem to cause some stress to the doe and is little used in the United Kingdom.

Conclusion
There is no perfect test for pregnancy in the goat. All of them have their drawbacks but they are useful in certain situations.

PROBLEMS ARISING DURING PREGNANCY

ABORTION

Abortion can occur for one of many reasons including infections, vitamin A deficiency or pregnancy toxaemia. The number of abortions that occur in the United Kingdom is probably fairly small, based upon the VIDA II records for the past six years (MAFF). The different kinds of infection that can be associated with abortion in goats are numerous and they include:

Enzootic abortion (chlamydia)
Toxoplasma
Salmonella spp
Mycoplasma
Vibrio
Q-fever (*Coxiella burnetii*)
Paratuberculosis (Johne's disease)
Listeria.

Procedure if Abortion Occurs
Isolate the doe, keep the placenta and aborted foetus and inform your veterinary surgeon. As a precaution, clean up and burn the material from where the abortion occurred, in case it was an infectious cause. Your vet will probably send samples to a laboratory for examination and the more of the placenta and foetus you keep, the better the chance of determining the cause. A blood sample from the doe will be required for many investigations. The percentage of examinations which reveal an infectious cause is fairly small but this is partly due to the fact that laboratories do not always receive the correct samples.

Confirmed Infectious Abortions
What can be done following diagnosed abortions in a herd depends upon what organism was involved. In some instances, it is best to mix all kidded or non-pregnant does with the aborted doe in order to try and *spread* the infection! The reason being that if the goats are infected while they are not pregnant, no disease will result but they will become immune and will not abort the following year. This policy should be adopted with chlamydial abortion (E.A.E.) and toxoplasma infections. With listeria and salmonella abortion one would strive to prevent the spread of infection. Your veterinary surgeon will advise you on this complex subject. One must remem-

ber that many of these infections cause abortion in sheep and goats and therefore when the species are run together, the risk of abortion is higher. Incidentally, rams will mate with does but the foetus is aborted.

Estimating the Age of the Aborted Kid

Sometimes it is useful to know at what stage of pregnancy the abortion occurred. This can be estimated from the size of the aborted kid. The following information may be of help:

length of foetus at 30 days: 1.4 cm
length of foetus at 145 days: 43.0 cm.

The eyes open between the 130th and 140th day of pregnancy.

CLOUDBURST (PSEUDOPREGNANCY, FALSE PREGNANCY)

Reports, surveys and personal experience indicate that this condition is quite commonly seen in goats in the United Kingdom. Why the condition occurs is unknown, but one theory is that it is more prevalent where does are not bred until they are 18-month-old goatlings. In other countries, where the practice is to mate the female at a younger age, this condition is less common. An abattoir survey in Scandinavia revealed three cases out of 1,020 goats. Two Indian surveys of a similar scale recorded a low frequency. By contrast, the author noted two cases in a much smaller sample of goats (231) in the United Kingdom.

Symptoms

Due to the influence of the ovarian hormone, progesterone, the doe's body becomes mistakenly convinced that it is pregnant. The signs and behaviour of pregnancy are seen and the goat can go to full term. At the time when kidding would have occurred the contents of the uterus (litres of fluid) are expelled, but no kid! Following the 'cloudburst' the doe may start to milk. Such does can be mated in subsequent years, can conceive and can even have a normal pregnancy. Cloudburst can be confirmed by the milk test. If diagnosed during pregnancy the condition can be terminated by your veterinary surgeon. This may enable the goatkeeper to have her served again that season.

Pregnancy Toxaemia
(*See* Chapter 4, page 69).

Prolapsed Vagina

The condition where the vagina of the pregnant doe starts to fall out through the vulva is seen occasionally. It is of little significance in itself but the doe may begin to strain as a natural reflex, and this is harmful. The delicate tissues also become prone to damage on hard objects and then bleed. The replacement of the vagina is a fairly simple job for your veterinary surgeon. He may either suture the sides of the vulva to prevent prolapse, or insert a plastic retainer sold for use in sheep. The condition is thought to be due to excessive abdominal pressure and is most likely to occur in overfat does. Just before kidding, the sutures or retainer should be removed.

Prevention
Prevention relies upon keeping does lean in the early part of pregnancy.

Preparing the Doe for Kidding

The doe in early pregnancy does not require too much food. Flushing, i.e., feeding a greater quantity of food just around mating, is sensible in order to produce a greater chance of conception. For the first three months of pregnancy the doe requires little more than maintenance, because the developing kid is only very tiny.

That does not mean that the quality of the food should be poor because all the vitamins, minerals, etc., must be available. Eight weeks before the expected kidding date, the ration should be increased by gradually feeding more concentrate. This ensures that the developing kid receives enough food and that the doe does not develop pregnancy toxaemia. The doe should be fed between 0.25 and 0.5 kilograms of concentrate just before kidding. More may be required if she is carrying triplets or quadruplets; how much, depends partly on the quality of the forage.

The Kidding

It is probably easiest to kid the doe inside, especially if lights and water are available. In order that the doe acclimatises to the micro-organisms in that environment, she is best placed there about fourteen days before the birth. This will ensure that her colostral antibodies are ideal for the kid. Exercise is good for her, and being out during the day is desirable. Signs of imminent kidding are a full udder, an enlarged slack vulva and restless behaviour. Most kiddings will proceed without any assistance from the owner and the best advice is to leave her alone to get on with it. Normal kiddings are over with fairly quickly.

After the water bag has appeared, discreetly check the doe every half-hour. If she strains without producing anything for longer than an hour, consult your vet for advice. Should the doe become exhausted and the kid's feet or head are visible, then you should help her.

First of all ensure that you have clean hands, washed in plenty of soap and warm water, and wash the doe's vulva. The second step is to identify what is presented at the vulva. If two feet are showing, feel further to identify either a head or a tail. Now you know which way the kid is presented. If only legs and no head or tail can be felt ring your vet for advice. If the kid is coming backwards or forwards, then you can proceed to pull it out. The pull should coincide with the doe's straining movements. Should you be confused about just what is presented at the vulva then do not just pull regardless because you may, for example, be pulling one leg of one kid and one leg of another. Similarly, it is easy to be confused and pull one hind leg and one fore leg and this will not come out either!

Reviving a Weak Kid
Some kids are weak when born, especially if they were born back-feet first, and they should be assisted. Hold them upside down by the back legs and swing them gently. This accomplishes two objectives; it clears the nostrils of any mucus or fluid and it stimulates their breathing. After swinging the kid, put your finger into the mouth to check for further mucus. Do not give up easily with a kid, many will respond if you keep stimulating them. Continue gently pumping their chests to encourage breathing.

Drugs to stimulate breathing can be given such as Respirot (Crotetham, Ciba-Geigy) or Dopram V (Doxapram, A.H. Robins). Both these preparations can be given by dropping them on to the kid's tongue; both are 'prescription-only' medicines.

Difficult Kidding (Dystocia)

A problem birth may arise from one of many causes. Examples are jamming of two kids at the pelvis, oversized single kids and kids awkwardly presented so that they cannot come out naturally. Some of these are illustrated in figs. 7.4 and 7.5.

Management of a Difficult Birth
What you do as a goatkeeper rather depends upon your experience

and your aptitude. Providing that you are suitably clean when investigating what is going wrong inside the doe, little harm will be done. Investigation is not the same as correction and, having discovered a real problem by feeling with your hand, the best course of action may be to summon your vet. If the kid has to be repositioned this has to be done with a great deal of skill and care, so as not to rupture the uterus. If you feel that professional assistance is required, call your vet in good time so as to give him a chance to deliver live kids. The more protracted the delivery, the less the chance of live kids.

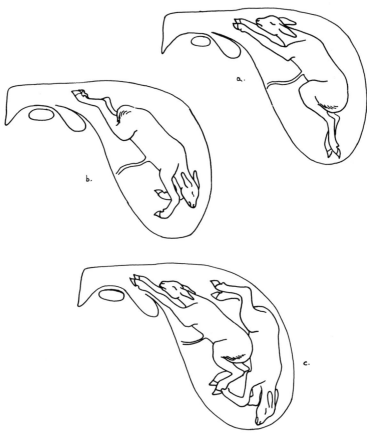

7.4 Normal presentations: (a) anterior
                         (b) posterior
                         (c) twins

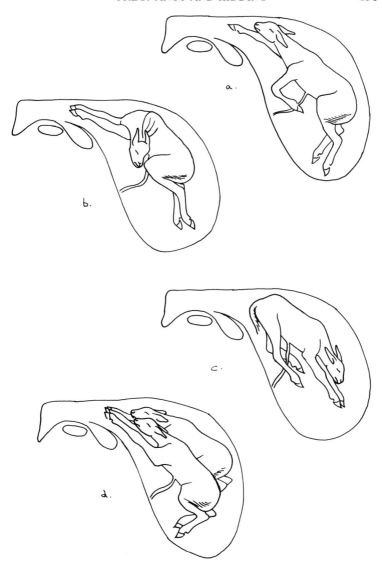

7.5 Abnormal presentations: (a) foreleg back
(b) head back
(c) breech
(d) twins jammed at pelvis

Ringwomb (failure of the cervix to dilate)
One reason for a protracted kidding is the failure of the cervix to dilate, making it impossible for the kid to be born. The doe strains but fails to produce anything at the vulva.

Recognising ringwomb
If a doe fails to produce anything after an hour of straining then she should be examined *per vaginum*. After a good wash in soapy water the owner may feel in to find out what is happening. Use plenty of soap as lubricant and slowly introduce the hand into the goat's vagina. If the hand and wrist can slide in easily, push further, attempting to identify a kid presented at the pelvis. If the fingers come up against the cervix no kid will be reached but the tube of the cervix identified. The 'ring' formed by the cervix will probably be open just enough to allow the operator to push two fingers through the hole.

Procedure
Frequently this problem can be overcome by a combination of drugs and patience. Drugs can be given which relax the cervix, allowing the kid to be born. This can be assisted by gentle help using one's fingers. The fingers are inserted into the cervix and gently pulled apart in an attempt to stretch the tissue of the cervix. If this treatment fails your veterinary surgeon will probably carry out a Caesarian section in order to deliver the kids.

Prevention
Do not allow does to become too fat before kidding and do encourage some exercise during pregnancy.

Oversized Kid (Foetus)
Sometimes a kid will be so large that it is impossible for it to pass through it's mothers pelvis. This is especially common in very small first kidders or in goatlings served unintentionally. In such instances, after straining for over an hour, the doe will fail to produce anything at the vulva. If the goatkeeper attempts a vaginal examination with a well-lubricated hand the condition can be easily recognised. Only the fingers will be able to pass into the vagina because the bones of the pelvis will prevent the operator from going any further. Call your veterinary surgeon whose help will be required to deliver the kid by Caesarian section. If the doe is worth very little one must always consider euthanasia.

1

2

A NORMAL KIDDING

1. The doe is lying down and straining. The front foot of the first kid appears at the vulva, still enclosed within the membranes.

2. After further effort by the doe, the nose of the kid now protrudes from the vulva and the second fore limb is just apparent. Note that the nose is presented before the crown of the head.

3. The head of the kid is almost completely clear of the vulva.

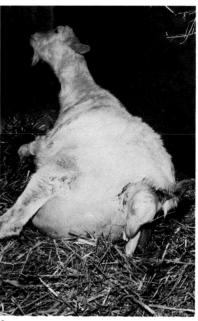

3

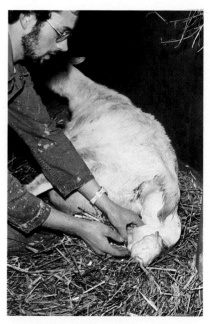

4. The head and neck are well clear, but more straining is required to push out the large shoulders and chest.

5. Assistance is being given to the doe because she now has to push out another large mass, the hips. Although it is not essential, goatkeepers could assist at this stage.

6. The first kid is delivered and the umbilical cord is being separated. If the goatkeeper wishes to assist, the cord is ruptured by pulling it apart (not cutting), at least 6 inches from the kid's navel.

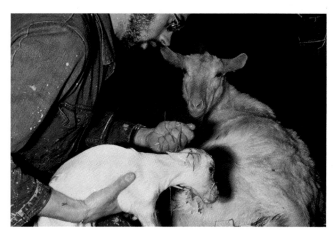

7. The newborn kid is quickly given to the dam.

8. The dam stimulates the kid by licking; she cleans off any fluids and membranes, assisting the drying of the kid's coat.

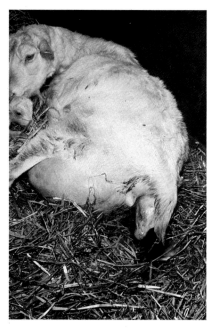

9. While the first kid is being revived the head of the second kid appears at the vulva. Normally a second kid is produced quite quickly. An interval of more than half an hour could be considered abnormal.

10. The second kid is quickly expelled while the dam concentrates on licking the first.

11. The dam attends to the second kid now that the first is well. The kid must be dried quickly to reduce heat loss. Goatkeepers should revive kids if the dam fails to do so.

12. The two kids are standing and sucking from the udder. The dam nuzzles and cleans them while she suckles.

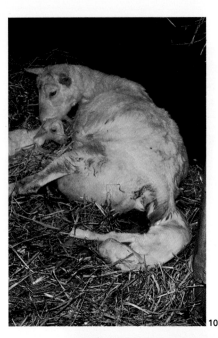

10

11

12

Deformed Kids

Abnormalities may be produced, such as kids with extra legs or two heads. They may be the cause of a difficult birth. Fortunately these kids often die soon after birth.

Caesarian Section

Some difficult births are impossible to correct without resort to a Caesarian operation. If a Caesarian is performed the doe is often less damaged than if the delivery is made via the birth canal. Often the chance of kids surviving is increased as well . Your veterinary surgeon will obviously decide when and how it should be carried out. Two basic methods can be used; in the first the approach is through the left-hand flank and the operation carried out under sedation and local anaesthesia. Alternatively a general anaesthetic may be given (Ketamine–Xylazine mixture) and the operation performed in the midline of the doe's abdomen. Another useful anaesthetic method is to induce with barbiturate and maintain with Halothane.

AFTER THE KIDDING

Ensure that doe and kids are comfortable and leave them alone as much as possible. Always leave the doe alone with her kids, never with other goats or the kids will become confused. From a discreet distance, try to observe if the kids drink colostrum. If kids do not suck within six hours assist them to do so. Dress the navel with a suitable antiseptic once the does and kids have established a good bond.

PROLAPSE OF THE UTERUS

Very occasionally, the doe will continue to strain after kidding and push out her uterus. You will see a large red ball of tissue protruding from the vulva, which may be bleeding. This is an **emergency** and the first thing to do is to call your veterinary surgeon. Do not attempt to do anything because chasing the doe around will probably result in this delicate organ being bruised and possibly torn. Whilst awaiting the vet, organise some buckets of hot and cold water, for washing purposes.

Treatment

Eversion of the uterus may result in bleeding, shock and infection. Your veterinary surgeon will quickly attempt to replace it and secure it into position again. He may give the doe an epidural anaesthetic, in an attempt to stop her straining, or he may simply raise the back legs of the doe in order to eliminate the straining. Infection he will control by placing pessaries (or tablets) directly into the uterus.

After-care
Providing the uterus is replaced within a few hours the doe should survive. The owner must be observant to make sure that the doe does not start straining again when the vet has left.

Retained Placenta (after-birth)

Normally the afterbirth is pushed out from the uterus, within three hours of the birth. Should the doe fail to cleanse herself within fourteen hours, then contact your veterinary surgeon. Do not just pull the afterbirth or it may break, leaving part of it in the uterus.

Goats that do not Kid on Time

If a doe goes over 154 days of pregnancy (count the day of service as day zero), it may be advisable to have her checked by your veterinary surgeon. There may be something preventing her having the kids, for example, a mild hypocalcaemia. The longer she goes on, the larger the kids grow. Large kids can present problems at birth and your vet may decide to induce the birth by injecting the doe with prostaglandin or a corticosteroid drug.

How Many Kids to Expect

One study of Saanen goats in the United Kingdom revealed that one kid is common in the first pregnancy, but the average is 1.4 kids. Does that had kidded previously gave birth to an average of 2 kids. The distribution is:

1 kid	20% of pregnancies
2 kids	60% of pregnancies
3 kids	18% of pregnancies
4 kids	2% of pregnancies
	100%

Very occasionally quins are produced (see plate 7.6).

PLATE 7.6
Quins born at the N.I.R.D.

Metritis

After the kidding the womb occasionally becomes infected and this is described as metritis. It may follow a retained placenta, a difficult kidding or just occur after a normal birth.

Symptoms
A discharge with colour ranging from dark brown through yellow can be seen coming from the doe's vulva. The goat may be quite sick and go off her food. It will be necessary to call your veterinary surgeon who may treat the condition in any number of ways. He may decide to put tablets into the uterus, if the neck of the womb has not closed, or he may decide to irrigate the uterus using a catheter. If the cervix has closed, injections of antibiotics are generally successful.

Mummified Kids

Occasionally a mummified kid will be produced. Mummification occurs after a kid has died in the womb, as a result, for example, of an infection. The one in plate 7.7 arrived two days after the birth of a live kid.

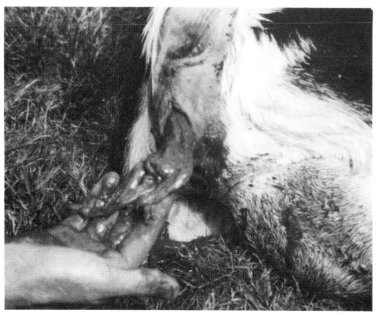

PLATE 7.7
Mummified kid

Chapter 8

OTHER CONDITIONS

GENERAL

ABSCESSES

Frequently found on goats, abscesses are local pockets of infection which contain pus. They tend to arise following small cuts which penetrate the skin. Germs entering at the same time as the cuts are responsible for the infection.

Symptoms
Swelling will be found below the skin anywhere on the goat's body. The problem is that blood-blisters, growths, cysts and other conditions all look and feel rather similar to the layman.

Treatment
First of all do nothing, wait and see if the suspected abscess bursts. Bathing it in warm salt water may assist the pointing process. If nothing happens over a period of several days get your veterinary surgeon to examine it. Never incise a lump unless you are absolutely sure what it is.

Your vet will check what the lump is, and incise it if necessary. He may give the goat an injection of antibiotic depending upon the severity of the condition.

CASEOUS LYMPHADENITIS (PSEUDOTUBERCULOSIS) (ABSCESSES)

This condition has not been recorded in the United Kingdom although it is prevalent in the United States.

ANTHRAX (*a notifiable disease*)

Anthrax is an acute fever produced by a germ *Bacillus anthracis* and generally resulting in sudden death of the animal. The same bacterium can cause disease in other animals such as cows, sheep and even humans.

157

PLATE 8.1
Abscess on nose of goat

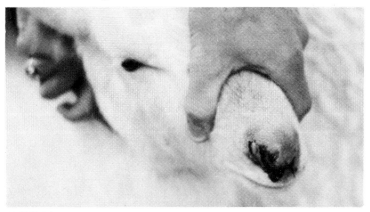

PLATE 8.2
Abscess after lancing

Symptoms
Animals are found dead or very sick with a temperature well above normal, for instance, 41°C (106°F).

Occurrence
This is very, very rare, in the United Kingdom because of the

regulations imposed by the Ministry of Agriculture. All stock-owners are obliged to report sudden deaths (cases where the owners see no signs of sickness prior to the death). These deaths are investigated by ministry staff, who check the dead animals' blood for the presence of anthrax bacteria. Positive cases are either burned or buried without being cut open. This is because the organism forms spores when in contact with the air. Spores are long-lived and can inhabit the soil for fifty years.

Treatment and control
Animals are rarely treated because death is so rapid. Should goatkeepers be suspicious, they are obliged to contact either their veterinary surgeon, the Ministry of Agriculture (Animal Health Division) or the police.

ARTHRITIS AND JOINT DISEASE

Few studies have been made of these problems in goats, and my experience of them is fairly limited.

Septic Arthritis
Arthritis may follow any generalised infection or bacteraemia (blood poisoning). The organisms settle in the joints, causing degenerative change and pain. Success with such cases depends upon how quickly treatment is started, but the outlook is often bleak.

Symptoms
The goat is very lame and swellings may be detected at the joints. Because of the pain resulting from the swellings the goats tend to lie down most of the time.

Treatment
Antibiotic injections may be prescribed by your veterinary surgeon but the damage is often so great that the goat fails to get better.

Prevention
Because this condition frequently follows an infection of the navel, cleanliness at birth is important. Dressing the navel of kids with iodine solution should prevent this type of infection.

Non-Septic Arthritis
This is a temporary problem resulting from an increased quantity of fluid in the joints.

Symptoms
Swelling of the joints, probably following twisting or sprain. The affected joint is swollen and painful but the goat is bright and continues to eat.

Treatment
Essentially rest, possibly supporting the joint with an elastic bandage to reduce further movement.

Viral Arthritis (C.A.E.V.)
In the United States arthritis in goats has been described associated with a retrovirus. At the time of going to press, it has not been isolated from goats in the United Kingdom. The virus has also been isolated from kids with disease of the brain and is called caprine arthritis — encephalitis virus (C.A.E.V.). In the United States the virus is widespread and antibody can be demonstrated in about 80 per cent of tested goats. The encephalitis form is generally seen in kids 2-4 months of age, and the symptoms are those of incoordination and paralysis. Arthritis is described in older kids and in adults. The symptoms are of swollen joints, sometimes becoming so bad that the joint no longer moves.

Control
There is no treatment for this condition. The transmission appears to be mainly from the dam at or around birth, especially in the colostrum. Infected does can be identified by testing blood samples.

<h2 style="text-align:center">AUJESZKY'S DISEASE (PSEUDORABIES)</h2>

Aujeszky's disease is a viral disease affecting primarily the pig, but occasionally causing disease in goats kept in close contact with pigs. When it does affect goats, it can be a very serious problem.

Symptoms
Unlike the situation in cattle, where the disease is also called the 'mad itch', no evidence of skin irritation is seen. The goats quickly develop symptoms of restlessness, sweating, screaming and finally paralysis. The course of the disease can be fairly rapid; some goats dying within twenty-four hours. In one outbreak in Holland, thirteen out of fifteen died in a space of ten days.

Control
The disease is rare in the United Kingdom, even in pigs. If a

premises is known to have the disease in the pig herd it would be sensible to keep the goats away from the pigs. It is possible that the disease may be eradicated from the pig population in the future, in which case the potential problem would disappear for goatkeepers.

BRUCELLOSIS

Brucella organisms infect an animal's placenta and udder, causing abortion and mastitis. A great deal of misunderstanding arises among goatkeepers on this topic, mainly stemming from the difference between the words possibility and probability. Perhaps I can clarify the situation.

Two strains of this organism infect goats, both of which can cause disease in humans; they are *Brucella melitensis* and *Brucella abortus*. *Brucella melitensis* is widespread in goats throughout the world but it has never been recorded in the United Kingdom.

Brucella abortus, which is a disease of cattle, can occasionally affect goats.* Having stated that it does affect goats, one must add that infection is extremely unlikely. The possibility of it causing problems in the future in the United Kingdom is even more remote because the disease has now been eradicated from cattle. However, goatkeepers should always be aware that brucellosis can be transmitted to humans by way of an infected animal's milk.

CANCERS IN GOATS (TUMORS, NEOPLASIA)

A cancer is the accumulation of tissue which results when cells suddenly start to multiply at a rate above normal. The reasons why they commence to multiply out of hand like this are largely unknown. We know that some of them are associated with viruses, but cancer as yet is largely a mystery. Probably the most comforting fact for goatkeepers to hear is that the frequency of cancers in goats is not particularly high.

The results of the two following studies puts the problem into perspective. Only fifteen out of 2,500 goats were affected in one United States survey and only seventy out of 800,000 in another abattoir survey. A second piece of information to remove concern from goatkeepers is that no one particular type of cancer seems to be a problem.

*see MATHUS, T.N. (1967) *Indian Journal of Veterinary Science*, 37, 272-86.

FOOT-AND-MOUTH DISEASE (*a notifiable disease*)

This highly contagious viral disease of all cloven-hoofed animal is not normally present in the United Kingdom. It is controlled by slaughter of the affected animals, compensation being paid by the Ministry of Agriculture. Because it spreads so rapidly, all owners must report suspicious symptoms immediately; I describe them below. If you suspect foot-and-mouth disease phone either your veterinary surgeon, Divisional Veterinary Officer or the police. Do not delay; there will be no charge made to you, even in the event of a false alarm.

Symptoms
Blisters or vesicles form in any of the following places: lips, tongue, teats, or the coronary band of the hoof. Affected animals tend to become lame and possibly salivate excessively. It is normally very mild, almost inapparent in the goat.

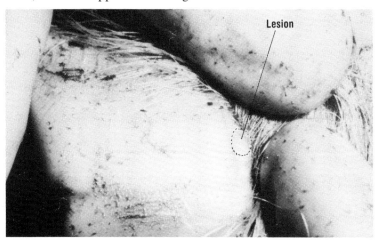

Lesion

PLATE 8.3
Foot and mouth disease — lesion on coronet

Control
The United Kingdom has a slaughter policy and treatment is never considered. The Ministry of Agriculture will deal with all aspects of the outbreak if it is confirmed.

Legal aspects of foot-and-mouth disease
This is a NOTIFIABLE DISEASE and owners are legally obliged to report any suspicious cases as soon as possible.

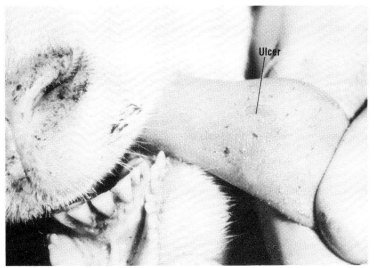

PLATE 8.4
Foot and mouth disease — tongue ulcer

LEPTOSPIROSIS

There are few reported cases of the disease that results from infection with leptospiral organisms. It can result in either a very acute disease following blood poisoning, or a mild form may be seen. In contrast to cattle the abortion form is not characteristic of goats, except as a complication of blood poisoning. A survey of goats in Morocco revealed that over a quarter of them had antibodies to the organism, suggesting that infection may be quite common. It is probably the case that sub-clinical leptospirosis is fairly widespread.

Spread of the disease
The organism is generally spread from one animal to another by urine, which often contaminates water supplies. The infection tends to settle in the kidney and thereby infects the urine. After infection, goats are known to harbour the bacterium in their kidneys for at least a month.

Symptoms
Most animals are found dead. Some show dullness and specific symptoms, occasionally with jaundice. Sometimes, after the blood-

poisoning phase the bacterium may remain in the nervous system and produce symptoms of encephalitis (*see* C.C.N., page 73.)

Treatment
In many cases discovery is too late for treatment but it may be beneficial to try and prevent carrier animals spreading the disease. Your veterinary surgeon will advise you on these aspects.

LISTERIOSIS (CIRCLING DISEASE)

Listeriosis is the disease caused by a bacterium *Listeria monocytogenes*. In goats it shows up in three forms, one affecting the brain, causing circling symptoms, the second being responsible for some cases of abortion. The third type is septicaemia or blood poisoning. It is not very common; for example, the Ministry of Agriculture laboratories only diagnose the problem in one or two samples each year.

Symptoms
Symptoms of the circling form are initially dullness, possibly with the goat head-pressing against an object. Later the animal tends to turn its head to one side and progresses to walk in circles. Over a period of a week, the affected goat becomes progressively weaker until it goes down, unable to rise. There may be a rise in temperature to about 40°C. Your veterinary surgeon will have to examine the goat in order to differentiate this disease from several others.

Abortions due to *Listeria* may reach 15 per cent of an affected group. They are reported to occur from three months of pregnancy onwards.

Spread of the disease
The organism is present in the soil. Infection can occur through the conjunctiva of the eye or from taking the organism in by mouth. There are thought to be special predisposing reasons why infection occurs and one of them is feeding silage. The reason for this is that the organism thrives in poorly made silage, especially if it is badly contaminated with soil. It is perhaps fortunate that few goatkeepers feed silage in the United Kingdom. However, soil may contaminate other foods if goats are badly managed and allowed to jump into hayracks and feeders.

Treatment
Poor response is obtained if animals have developed the circling

form of the disease but antibiotics such as penicillin or chlortetracy-cline may be tried. Once symptoms are noted in a herd, it would be sensible to check the in-contact animals for rises in body temperature. They would probably benefit from preventive antibiotic treatment. In one outbreak, affecting fourteen adults and two kids, two adults and two kids died, the other adults responding to treatment with chlortetracycline.

Prevention
Avoiding the spoilage of food by soil, especially if silage is fed, is important. Silage must be well made so as to ensure complete fermentation in order to prevent the organism multiplying in it.

MAEDI-VISNA (M.V.)

This chronic pneumonia of sheep was recently diagnosed for the first time in the United Kingdom; maedi-visna can also infect goats.

Symptoms
This disease is caused by a virus but it has an extremely long incubation period of at least two years. The symptoms are described as 'progressive', very slowly becoming more apparent. Listlessness and loss of weight become obvious and the animal breathes more rapidly than normal. Goats showing nervous symptoms have also been reported.

Treatment and control
There is no treatment for this condition and slaughter of affected goats is the only answer. Recently (January 1982) the Ministry of Agriculture has introduced a voluntary scheme for maedi-visna-free herds. This is aimed primarily at sheep flocks but goatkeepers are welcome to take part. Blood samples are taken from goats and tested at a laboratory, although the owner must pay for the cost of the qualifying test. Contact your Divisional Veterinary Officer (D.V.O.) for further information. The D.V.O. can be contacted by telephone; his number appears in the directory under 'Agriculture, Ministry of'.

SCRAPIE

This unusual disease occurs in sheep and extremely rarely in goats. It is caused by a minute particle smaller than a virus and gives symptoms of an abnormal gait and itchiness. Affected animals rub and bite themselves, progressively losing weight. Goats become infected from infected sheep. Scrapie is incurable and affected animals should be slaughtered. There is no vaccine available to control this disease.

TUBERCULOSIS

Goats can occasionally contract tuberculosis. The source of infection used to be mainly from tuberculous cattle. As the prevalence of T.B. in cattle is now infinitesimal, infection in goats is almost unknown in the United Kingdom. Cattle are tested in order to prevent tuberculous milk being consumed by humans, and goats running with cattle should be tested at the same time. In the author's opinion all goats producing milk for human consumption should be tested (there is no legal obligation to do this). Goats are generally infected with the avian strain of T.B. but, very occasionally, the bovine strain has been recorded. A few cases of T.B. in goats have been found recently in the United Kingdom.

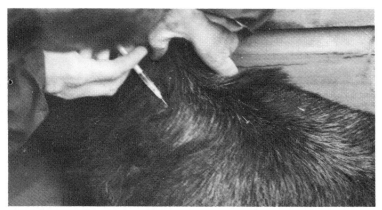

PLATE 8.5
Injecting tuberculin into the skin for a tuberculin test

URINARY CALCULI (STONES IN THE BLADDER)

Male goats are especially prone to blockage of the urethra by 'stones' which collect in the bladder. If they leave the bladder and pass into the narrow penis, they cause obstruction and blockage to urine flow.

Symptoms
The goat rapidly becomes distressed and repeatedly attempts to urinate; crystals may be noticed on the belly hair. Place the goat on a

dry concrete surface and observe to see if he can pass any urine. Sometimes the pressure becomes so great that the bladder bursts and urine floods into the abdominal cavity. This may result in the goat becoming brighter for a few days because the pain associated with the blockage disappears.

Treatment
Sometimes the obstruction can be relieved by surgery; I find only about half of the cases are successful. The operation would not normally be performed on a breeding male because it may affect his fertility.

Prevention
The main predisposing cause of this condition is dirty drinking water buckets. Goats will not drink dirty water and if their water is not clean they will ration themselves. Stall-fed males on high concentrate/low roughage diet are particularly at risk. Decreasing the concentrate may help to avoid the problem. It may also be prudent to avoid concentrates that are high in calcium.

Plate 8.6 illustrates the type of stones (or calculi) found in the bladder of a goat that was suffering from this problem.

PLATE 8.6
Calculi from the bladder of a goat suffering from urinary obstruction

MIDDLE EAR DISEASE

Infection of the middle ear may cause goats to go round in circles as a result of disturbances to the balance mechanism (*see* plate 8.7).

PLATE 8.7
Goat with middle ear disease

Treatment
Treatment with antibiotics can restore the goat to normality within a
few days. The chances of recovery are good. Also it is beneficial to
assist the goat to eat and drink until it is better. The symptoms are
very similar to those of several other diseases such as Gid, C.C.N.
and listeriosis. Your veterinary surgeon will advise you on this.

SKIN DISEASES

Some diseases causing damage to the skin have been described in
other chapters. *See* ORF, page 103, and also pages 107 and 162.

RINGWORM (DERMATOMYCOSES)

Despite the common name, this disease is a fungal condition and
nothing to do with worms: Ringworm can affect all animals and it
can be transmitted to humans. Normally a disease of the winter
months, it is especially prevalent when animals are housed together.
In goats several fungi can grow on the hair and skin; they include
Trichophyton verrucosum.
    Ringworm does relatively little harm to the goat, but it looks
unsightly and it is a potential danger to humans.

Symptoms

A grey-white crusty appearance to small areas of skin should alert the goatkeeper to the possibility of ringworm infection. The skin is usually thickened and the hairs thin or absent. There is generally no itching or evidence of irritation. Enlargement of the affected areas occurs over a period of weeks and different patches may join up to form one large area. The fungus in the middle tends to die, due to lack of oxygen and this accounts for the descriptive name, ringworm.

PLATE 8.8
Ringworm

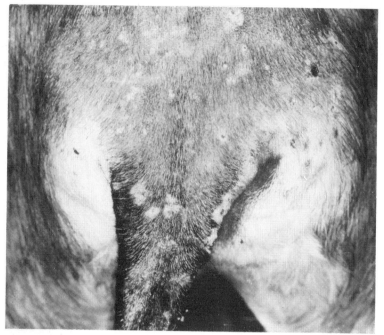

PLATE 8.9
Ringworm

The predisposing causes
Warm, humid conditions are contributory to the growth of fungus on the hair and skin. Badly ventilated goathouses are ideal for its spread. Young animals tend to be more susceptible to ringworm than older animals. Spread of the disease is either by direct contact or from contact with contaminated objects.

Treatment
Very effective treatments are available these days and they can be used either locally on the skin or given by mouth. Fungicidal preparations are applied as a liquid dressing on to the affected areas. Griseofulvin is an antifungal drug which inhibits the fungal growth on the hair shaft. It is generally given to the goats mixed in with their concentrate ration.

Prevention
The disease is best prevented by avoiding humid housing. Avoiding

contact with contaminated objects can sometimes be difficult, even in the best-managed animals.

MANGES

Infestation of the skin by parasitic mites is referred to as mange. There are several types of mange which affect the goat, four of them occur in the United Kingdom:

- Chorioptic mange;
- Demodectic mange;
- Psoroptic mange;
- Sarcoptic mange.

Although they are all caused by mites, it is important to know which one is affecting the goat. They are of varying significance, some causing only mild problems, others being virtually incurable. For this reason, your veterinary surgeon will probably take a sample for laboratory examination, in order to identify the mange involved.

Chorioptic Mange (Heel or Leg Mange)

The mites, *Chorioptes caprae*, infest the skin of the lower leg of goats. They are responsible for a little irritation and they cause the goat to look unsightly. Housed goats in the winter are more likely to suffer from the problem than grazing animals.

Symptoms and treatment

Itchiness may be noticed and there may be small crusty scabs. It mainly affects the heel region. Treatment with a variety of organophosphorus compounds will relieve the symptoms. In small herds these can be appled directly to the skin with a paint brush, but you must be careful to cover all of the legs. A repeat treatment two weeks later would be sensible.

Demodectic Mange

The mites, *Demodex caprae*, invade the hair follicles and sebaceous glands of the skin. This causes a chronic inflammation and the development of small pustules or abscesses. The disease spreads slowly on the skin of affected goats. It does not spread rapidly from goat to goat. In France the problem is increasing, and it is thought that the keeping of goats in intensive conditions is conducive to the spread of the mange. Goatlings between the ages of 10 and 15 months are mainly affected.

Symptoms
Small lumps are noticed in the skin, varying in size from a match head up to a small pea. They may be like a cyst or 'bag of fluid'. Little irritation is noticed, probably because the mite is so deep-seated. The areas mostly affected are at the front of the animal, neck, shoulder and chest. The animals can become so severely affected that they eventually die.

Treatment
Response to treatment is generally poor. Treatment with drugs effective against mites is frequently unsuccessful. Malathion is an example of one such compound. Some authors suggest that the pustules be cut open and then painted with tincture of iodine.

Psoroptic Mange
This mite is found mainly in the ears of affected goats where it causes some irritation and head shaking. It may also extend to the poll and even to to the legs. The variety of mite which causes sheep scab is distinct from *Psoroptes communis caprae* which affects only goats. Gamme Benzene Hexachloride and Gammexane have both been successfully used in the treatment of this mange.

Sarcoptic Mange
A serious mange causing crusty scabs on the skin is due to the mite *Sarcoptes scabei*. This form of mange appears to be more serious in goats than in other species of animals. The mites burrow in the skin and lay their eggs in tunnels. After hatching, the nymphs work their way to the surface of the skin where they become adults.

Symptoms
Scabby patches on the skin, especially on the head and neck, causing itchiness, are characteristic of the disease. Affected animals lose condition and appear very unsightly.

Treatment
Results are frequently disappointing although success has been claimed using 'Neguvon' (Bayer Chemicals, Germany) as a weekly wash, at a concentration of 0.2 per cent. Other cases have had to be destroyed after several months of unsuccessful treatment. Repeat washes with the acaricide (dip) have to be carried out because the eggs of the mite are not destroyed. Thus the repeat wash tends to kill the mites after they have hatched.

PLATE 8.10
Sarcoptic mange

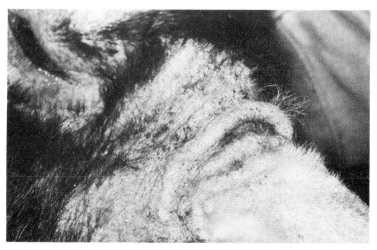

PLATE 8.11
Sarcoptic mange

Control
The microscopic mite can be spread from goat to goat either by direct contact or by equipment. Care should be taken to avoid using

the same brushes, feeding pails and similar equipment for both infected and clean animals.

LICE INFESTATION

Lice are found in the coat and they are just visible to the naked eye. Infestation with lice causes intense irritation and itching, usually during the winter months. Louse powder will normally control the problem. It is possible to eradicate the parasite from goat herds by treatment with such preparations as coumaphos (Asuntol sheep dip, Bayer) on three occasions, separated by ten-day intervals. Autumn is the optimum time for treatment. Goats can be either dipped or sprayed but in either case scrubbing to assist good skin penetration will help to ensure eradication is achieved. Lactating goats should be treated immediately after milking, in which case milk need not be discarded. The louse spends all its life on the host and does not infest the house or surroundings.

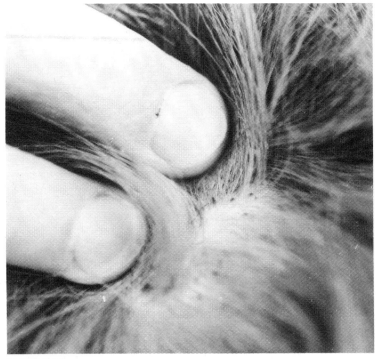

PLATE 8.12
Lice in the coat

PLATE 8.13
Lice infestation — general appearance

HARVEST MITE INFESTATIONS

Infestations with this mite causes irritation and rubbing. It is mainly seen as greasy areas on the legs and lower body. The infestation commences in August in the United Kingdom. The mites are red and can be seen with the aid of a magnifying glass.

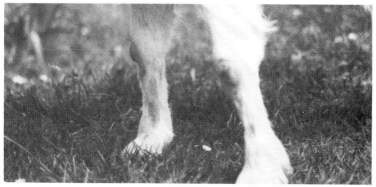

PLATE 8.14
Harvest mite infestation

Treatment
Remove the goat from the pasture and dress the areas with Benzene Hexachloride. If possible, the goats should be returned to a different pasture.

LUMPY WOOL (DERMATOPHILUS)

Damp, humid conditions can predispose to a coat condition described as a 'paint brush' appearance. Infection of the superficial layers of skin and hair by an actinomycete (a minute fungus) is responsible. It is normally seen during the winter months.

Treatment
Antifungal drugs and Streptomycin can be used but it is probably better to wait for improved weather conditions.

Prevention
Avoid the predisposing conditions of damp and humidity.

Chapter 9

ACCIDENTS, EMERGENCIES AND POISONING

ACCIDENTS AND EMERGENCIES

These types of problems fall into two categories, firstly those that the goatkeeper can treat and secondly those for which first-aid treatment should be given before your veterinary surgeon arrives. If in doubt, carry out the first aid and then ring your vet for his opinion as to what needs to be done. For quick reference, I shall list the topics that I shall cover in this chapter.

1. Control of bleeding
2. Tear wounds (lacerations)
3. Choke
4. Bloat
5. Electric shock
6. Snake bite
7. Fractures
8. Stab wounds
9. Objects in the eye

CONTROL OF BLEEDING

How to manage this problem depends upon where it is on the goat. If the bleeding is somewhere on a limb it is easier to deal with than bleeding from other places because a tourniquet can be appled for a short time. If the bleeding is severe, then although one should be as hygienic as possible, the immediate danger to the animal is loss of blood. In these instances, the first-aider is obliged to plunge in and do something quickly.

Arresting Bleeding from a Limb

For a tourniquet, a piece of stick and a length of string or baling twine are required. The twine is looped around the leg and wound over the stick so that the stick acts as a lever tightening the twine when turned.

177

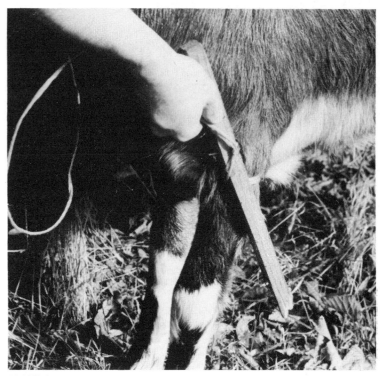

PLATE 9.1
Applying a tourniquet using baling twine and a stick

Never leave the tourniquet on for long, otherwise irreversible damage can result. In a case of severe bleeding apply a tourniquet, then ring your vet; return and release the tourniquet from time to time (every fifteen minutes) until he arrives. If the haemorrhage is not very severe, clean the wound with cotton wool soaked in dilute antiseptic (e.g. Dettol), clip the hair all round the wound, bathe the wound again and then apply a dressing. The dressing consists of bandage or clean (old) cotton sheets. If you have an antiseptic dusting powder, apply that before bandaging. Finally, release the tourniquet.

Control of Bleeding from Elsewhere
This is more difficult and the principle is to apply direct pressure using a clean material such as an old cotton sheet. If severe, your veterinary surgeon will have to deal with it, probably by stitching. A

small wound may be temporarily plugged with a piece of cotton
wool.

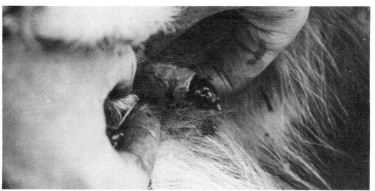

PLATE 9.2
Bleeding from penis resulting from attempting to leap over a barrier

Tear Wounds or Lacerations

If the bleeding is not severe then these may be treated by the owner
provided that the goat has been vaccinated against tetanus. Clean
up the area with cold water and clip away hairs from the edge of the
wound. Use tweezers to remove any foreign bodies such as splinters
of wood. Irrigate the wound with dilute antiseptic such as Dettol.
Allow it to dry and then apply antiseptic powder. Smear Vaseline
around the wound and apply a clean dressing of gauze and bandage.

In hot weather, a fly-repellant should be used, preferably one
combined with antiseptic powder such as Negasunt (Bayer).

Choke

Obstruction of the oesophagus (gullet) in the ruminant is of much
more consequence than a similar problem in, say, a human. The
reason of course is that gas is constantly being produced in the
rumen and it has to get out, otherwise the rumen fills up like a
balloon (bloat).

Pieces of apple or carrot, for example, may lodge in the oesophagus. Sometimes the bulge can be seen or felt on the left-hand side of
the neck near the windpipe. If possible, try and milk the object
down into the stomach by applying massage to the neck.

If this fails, let your veterinary surgeon deal with it. Should the
goat be in a very distressed state, puncture the rumen (*see* Bloat,
below).

Bloat Emergency Treatment

Only carry out the following if the goat is in danger of death; otherwise refer to Bloat, page 109.

PLATE 9.3
Correct site for the emergency relief of bloat

Take a very sharp knife and stab the left-hand flank in the area indicated in plate 9.3. When a goat is bloated the area which can be stabbed is quite large. It will be fairly obvious because it is soft when pressed. Put the knife in well behind the rib cage and fairly high up, because the gas will be at the top of the rumen. A fairly 'high' stab prevents leakage of rumen contents into the peritoneum.

It is not an easy thing to do, because the stomach behaves as if it were a balloon. Remove the knife and the gas will escape rapidly.

ELECTRIC SHOCK

Switch off the current before touching the goat! If you think that she is still alive stimulate her by massage. Keep her warm using plenty of straw or a rug over her back.

Prevention
Keep all electrical fittings high, out of the range of goats.

SNAKE BITES (ADDERS)

Occasionally to be found in goats, it is generally seen at the front end, particularly the lips, muzzle and neck. Swelling is noted; the goat may appear shocked and two fang marks may be detected. The swelling is very painful and causes concern to the goat.

Treatment
It is probably wise to get your veterinary surgeon to examine the bitten goat because the effects can be fatal. If the bite is on the leg, then the application of a tourniquet is advisable (for a short period) in an attempt to localise the toxin. Antibiotics are useful because many of the bites are infected with the germs of tetanus and gangrene.

FRACTURES

I have only seen fractures of the forelegs, usually those are of kids, trapped between horizontal bars. Active kids may attempt to jump from one pen into another and in doing so, sometimes catch their legs. Pen walls should always be sufficiently high to prevent this occurring.

PLATE 9.4
Fracture below the knee

First aid

Leave the goat quiet and summon your veterinary surgeon. You could attempt a rough splint as a temporary measure, but most animals with fractures will lie quiet and still. An adult goat should be propped on her brisket with bales in order to avoid bloat developing. It may be necessary for someone to stay with the goat in order to give her support.

Treatment

This very much depends upon where the fracture is; those below the knee or hock can be treated by external support, splint or plaster

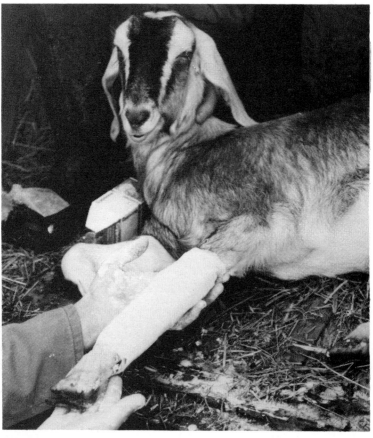

PLATE 9.5
Applying the plaster cast

cast. Breaks high up on the leg may only be repaired by surgical technique and the goat would have to be extremely valuable to warrant the expense.

Stab Wounds

Wounds that result from a spike or similar object penetrating the skin and muscles are potentially very dangerous. Do not be lulled into a false sense of security by the small size of the hole. The reason, of course, is that any infection that may have entered tends to become trapped inside. It cannot be washed and cleaned from the outside and thus one normally has to rely on antibiotic injections. Tetanus is the most serious result of this type of wound, and if any of your goats suffer such an accident obtain professional help quickly.

Colic

The pain that results from digestive upsets can be very severe and the goat will appear very distressed. Fortunately the pain is generally spasmodic and will pass off fairly quickly. Keeping the goat occupied by walking her will normally be enough treatment until the pain passes.If the problem persists your veterinary surgeon will probably give the goat a smooth muscle relaxant drug such as Buscospan (Boehringer Ingelheim) or Isaverin (Bayer).

Foreign Body in the Eye

Hay seeds may fall into the goat's eye and require to be removed. If it cannot be readily pulled from the eye, use a little sugar to stimulate tears. Failing this, touch the object lightly with a greased brush to pick it out.

POISONING

Lead Poisoning

There are very few reports of lead poisoning in goats, although there are several descriptions of experimental poisoning.

Lead is fairly commonly found around farms and homes, being present in paint, putty, batteries and felt. Goats are less inclined to lick lead objects but they may chew at wood and take in lead at the same time.

Symptoms
The layman will be unable to distinguish the symptoms from those of many other diseases such as C.C.N. (page 73). Blindness, inco-ordination, head-pressing and teeth-grinding are all seen.

Treatment
Some goats will respond to treatment but many will be too far gone and die. Injections of a substance known as calcium versenate will be given by your veterinary surgeon into the goat's vein. Drenching the goat with magnesium sulphate will help prevent the further absorption of lead from the intestine. The goatkeeper could drench the goat with two teaspoonfuls of Epsom salts dissolved in water while waiting for the veterinary surgeon to arrive. Because the symptoms are identical to several other diseases showing similar nervous symptoms, laboratory tests may have to be carried out. From a live animal, a blood sample will be collected, but if one animal from a group has died, then the kidneys will be collected from it at post-mortem examination.

Prevention
Always try to prevent goats coming into contact with sources of lead, especially lead paint.

Fluoride Poisoning
See FLUOROSIS, page 101.

PENTACHLOROPHENOL POISONING

Goats can succumb to poisoning with this substance, used abundantly as a timber preservative. Do not allow goats access to timber treated in this way until it is thoroughly dry.

Symptoms
Convulsions and other nervous symptoms followed by death.

Treatment
There is no specific antidote and the prospect is extremely poor. Your veterinary surgeon may try using non-specific treatment but there is nothing that will help much.

DIESEL FUEL POISONING

Poisoning with this substance is reported to give rise to dullness, pneumonia and nervous symptoms.

PLANT POISONING

There are a tremendous range of plants that can poison goats, many of them giving rise to a wide range of symptoms. In my experience, only a limited number of them are commonly seen and these I shall cover in detail. It is outside the scope of this book to deal with all the other possible causes of poisoning and the reader could refer to the HMSO publication number 161 for further information. Some general comments about plant poisoning in goats are worth while noting.

The Quantity of Plant Eaten

If only small quantities of dangerous plants are eaten, then the effects may be very slight. This is especially true if the goat has a rumen full of 'safe' food such as hay or grass. In this case the poison is 'diluted' in the rumen and has less effect.

The Husbandry System

When given a choice, goats on extensive grazing or browse tend to avoid poisonous plants. This is true of course until food becomes scarce in which case the goats may be forced to eat poisonous plants. If goats which are accustomed to being yarded are given the rare chance to browse they may take the opportunity to try anything within their reach and perhaps poison themselves.

RHODODENDRON POISONING

This is the most common type of poisoning that I see; even a few leaves seem to cause the symptoms to develop. The toxin involved is andromedotoxin.

Symptoms
Characteristically, the goats retch, vomit, salivate and become very depressed. Sometimes they are ill for several days and develop laboured breathing.

Treatment
Ephidrine may be given at a dose rate of 1 mg/kg of body weight, in order to counteract the effect of the toxin upon the heart. Vitamin injections are very useful in order to speed up the detoxification of the poison. If your goat is not vaccinated against enterotoxaemia your vet will probably prescribe a penicillin injection because of the

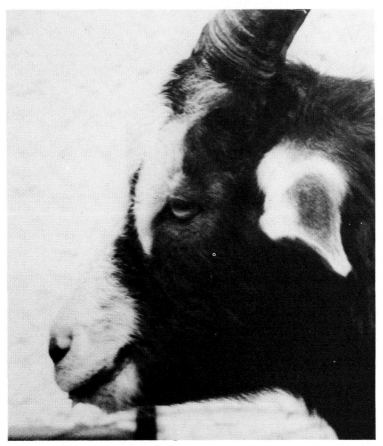

PLATE 9.6
Goat suffering from rhododendron poisoning (vomit on lips)

poison's effect on the digestive system. If the goat is valuable, your vet may perform an operation to remove all the stomach contents, via the left flank.

First aid treatment
Stimulants such as alcohol may be administered; for example, diluted whisky. Drenching with tea may also be attempted while you are waiting for the veterinary surgeon. Drenching is not recommended if the goat is still retching badly because suffocation may result.

RAGWORT POISONING

Ragwort is common in the United Kingdom and causes severe damage to the liver. The damage is gradual and irreversible, so there is no treatment for affected animals. The symptoms reflect the damage to the liver and include loss of condition, poor appetite and anaemia. In very severe cases the eyes and mouth become yellow due to the development of jaundice. Hay containing dried ragwort is still dangerous to animals.

KALE POISONING

Kale or rape is frequently fed to cattle and sometimes to goats. Excessive quantities can result in damage to the red blood cells which rupture. This results in the breakdown products of blood being lost in the urine. One of the symptoms of poisoning is, therefore, red urine. Other symptoms which may follow are weakness and anaemia. If kale poisoning develops, the only treatment normally required is to move the goats off the kale and on to other forages such as grass or hay.

OXALATE POISONING

Plants such as sugar beet tops or rhubarb can give rise to symptoms of hypocalcaemia (see MILK FEVER, page 112). Treatment involves the administration of calcium borogluconate in order to reverse this state.

OAK LEAF POISONING

Excessive feeding of oak leaves may lead to damage to the bone marrow and subsequently anaemia. The bone marrow is essential for producing red blood cells and if it is damaged fewer red blood cells are manufactured, which leads to anaemia. Oak leaves fed in moderation are fairly harmless.

FRUIT TREE LEAF POISONING *(Prunus)*

The leaves of the Prunus family are poisonous to goats because they contain a cyanogenetic glycoside called amygdalin. When fresh, the leaves are harmless but when dry and wilted they contain hydrogen cyanide. This compound combines with the oxygen-carrying structure of the red blood cell making it unable to carry oxygen. The symptoms are those of lack of oxygen with bright cherry red colour to the mouth and other membranes. Treatment involves the administration of sodium nitrate and sodium thiosulphate into the vein. Substances to stimulate breathing would help but the course of the disease may be too rapid for treatment to be given.

REFERENCES

JOURNALS

British Goat Society Journal.
Dairy Goat Journal, Scottsdale. Arizona. USA.
Farmers Weekly, London.
Goat Veterinary Society Journal, BVA, London.
International Sheep and Goat Research, Scottsdale, Arizona, USA.
Journal of the American Veterinary Medical Association.
Symposium of the American Sheep and Goat Practitioners (1976).
The Veterinary Record, BVA, London.

Also world-wide selection of journals and abstracts.

BOOKS

Blood, D.C., Henderson, J.A. (1974), *Veterinary Medicine,* Baillière, 4th edn, London.
British Poisonous Plants (1954), HMSO, London, Bulletin 161.
Duphar Vitamin Guide for Feedmen and Veterinarians, Duphar. Amsterdam.
Gall, C. (1981), *Goat Production,* Academic Press, London.
Guss, S.B. (1977), *Management and Diseases of Dairy Goats,* Dairy Goat Publishing Corporation, Scottsdale, Arizona, USA.
Hetherington, L. (1979), *All About Goats,* 2nd Edn, Farming Press, Ipswich.
Mackenzie, D. (1980), *Goat Husbandry.* 4th Edn, Faber, London.
Merck Veterinary Annual (1973), 4th Edn, Merck and Co., Rahaway. N.J., USA.
Theil, C.C., and Dodd, F.H., (editors) (1977), *Machine Milking*, N.I.R.D., Reading.

INDEX

191